Hydropower Development in the Mekong Region

The Mekong Basin is home to some 70 million people, for whom this great river is a source of livelihoods, the basis for their ecosystems and a foundation of their economies. But the Mekong is also currently undergoing enormous social, economic and ecological change, of which hydropower development is a significant driver. This book provides a basin-wide analysis of political, socio-economic and environmental perspectives of hydropower development in the Mekong Basin. It includes chapters from China, Thailand, Laos, Cambodia and Vietnam.

Written by regional experts from some of the area's leading research institutions, the book provides a holistic analysis of the shifting socio-political contexts within which hydropower is framed, legitimized and executed. Drawing heavily on political ecologies and political economics to examine the economic, social, political and ecological drivers of hydropower, the book's basin-wide approach illuminates how hydropower development and its benefits and impacts are linked multilaterally across the basin.

The research in the book is derived from empirical research conducted in 2012–2013 as part of the CGIAR Challenge Program on Water and Food's 'Improving Hydropower Decision-Making Processes in the Mekong' project.

Nathanial Matthews is Global Research Coordinator on the CGIAR Research Program on Water, Land and Ecosystems. Previously, he was a lecturer in the Department of Geography at King's College London, UK.

Kim Geheb is the Mekong Regional Coordinator for the CGIAR Research Program on Water, Land and Ecosystems, and is based in Vientiane, Laos PDR.

Earthscan Studies in Water Resource Management

For more information and to view forthcoming titles in this series, please visit the
Routledge website: www.routledge.com/books/series/ECWRM/

Hydropower Development in the Mekong Region

Political, socio-economic and environmental perspectives

Edited by Nathanial Matthews and Kim Geheb

LONDON AND NEW YORK

First published 2015 by Routledge

2 Park Square, Milton Park, Abingdon, Oxfordshire OX14 4RN

711 Third Avenue, New York, NY 10017

Routledge is an imprint of the Taylor & Francis Group, an informa business

First issued in paperback 2018

British Library Cataloguing-in-Publication Data
A catalogue record for this book is available from the British Library

Library of Congress Cataloging-in-Publication Data
Hydropower development in the Mekong Region : political, socio-economic, and environmental perspectives / edited by Nathanial Matthews and Kim Geheb.
    pages cm – (Earthscan studies in water resource management)
    Includes bibliographical references and index.
    1. Water resources development – Mekong River Watershed.
    2. Water-power – Mekong River Watershed. I. Matthews, Nathanial.
    II. Geheb, Kim.
    TC513.M45H94 2015
    333.91´4150959–dc23   2014020370
ISBN: 978-0-415-71913-1 (hbk)
ISBN: 978-1-138-37750-9 (pbk)

Typeset in Bembo
by HWA Text and Data Management, London

Contents

Figures

Tables

Contributors

Nga Dao is the Founder and Executive Director of the Center for Water Recourses Conservation and Development (WARECOD), Vietnam, and adjunct lecturer in York University's International Development Studies Program, Toronto, Canada. She is also the founder and member of the Management Board of the Vietnam Rivers Network. Nga Dao has worked and conducted research in southeast Asia for more than 20 years. Her research focuses on the issues of water governance, development-induced displacement and agrarian change in the region.

Kim Geheb is a geographer and holds a doctorate awarded by the School of African and Asian Studies at the University of Sussex, UK. He is currently the Mekong Regional Coordinator for the CGIAR Research Program on Water, Land and Ecosystems, and is based in Vientiane, Laos PDR. Kim Geheb has 16 years of experience conducting, implementing, managing and leading natural resources research-for-development projects in the developing world. He draws his experience from Kenya, Uganda, Tanzania, Ethiopia, Laos, Cambodia, Thailand, China and Vietnam, and has conducted, designed and implemented initiatives in river basins and catchments, water resources, fisheries, soils, livestock, land and protected areas. He is a fellow of the Royal Geographical Society and the Royal Society for the Encouragement of the Arts, Manufactures and Commerce.

Kimkong Ham is a senior lecturer and researcher and the head of the Research Unit at Department of Natural Resources and Development, Royal University of Phnom Penh, Cambodia. He specializes in environmental economics, water governance, the political ecology of hydropower development, livelihoods, climate communication-related to risks and climate change adaptation. Kimkong Ham is currently involved in a number of research projects such as examining the politics of the Lower Sesan 2 Dam in Cambodia, communicating water-related risks to improve local adaptation in the Mekong Delta and examining climate change and resilience on land use and water resources that affect farming communities' livelihoods and food security in the Stung Chreybak watershed in the Tonle Sap Lake in Cambodia. He has been involved in multi-disciplinary research projects, which aim at capacity building and supporting policy development and practices with local and national government agencies and academia.

Samchan Hay is a project manager at E&A Consultant Co., Ltd in Cambodia. Previously he was a research associate in the Department of Environmental Science at Royal University of Phnom Penh, Cambodia, involved in the 'Improving Hydropower Decision-Making Processes in the Mekong Basin' project of the Challenge Program on Water and Food. His research interest is in environmental impact assessment, political economy, development policy and natural resource management. He has more than ten years of experience working with communities, local NGOs and international aid agencies.

Xing Lyu is Associate Professor in International Environment Politics at the Southeast Asian Institute, Yunnan University in Kunming, China. He is also currently part of the core team of the Mekong Research Fellowship Program that fosters the development of water professionals in the Mekong Region. Xing Lyu has engaged with international, regional and national organizations on water governance, cross-border trade and investment, regional integration and teaching across the Mekong Region. His recent research interests focus on implications of regionalization and impacts of knowledge production and legitimization on decision-making processes, especially in the environment field. He started his career as practitioner and then trainer for community development and environmental conservation projects.

Nathanial Matthews is a political and environmental scientist whose principal interests lie in understanding and mitigating the impacts of political and economic change on people and the environment in developing countries, especially concerning the interconnected challenges across water, energy and food. From 2012 to 2014, he was the principal investigator on the 'Improving Hydropower Decision-Making Processes in the Mekong' project that guided this book. Nate has professional and fieldwork experience spanning 14 countries around the globe primarily focused in Asia and Africa. He has consulted on a range of research for development (R4D) projects with agencies such as: Oxfam, SIWI, UNEP and UNESCO. Previously, he was the education and research officer at the International Water Centre and a lecturer at King's College London, UK. He currently works as Global Research Coordinator on the CGIAR Research Program on Water, Land and Ecosystems.

Carl Middleton is a lecturer on the MA in International Development Studies (MAIDS) Program in the Faculty of Political Science, Chulalongkorn University, Thailand. He graduated from the University of Manchester, UK, with a bachelors in civil engineering, and a doctorate in environmental chemistry. Before joining the MAIDS program in 2009, he spent seven years working with international and local civil society organizations throughout the Mekong Region. His research interests are the politics and policy of the environment in Southeast Asia, with a particular focus on water and energy governance.

Naho Mirumachi is lecturer in the Department of Geography, King's College London, UK. She has over ten years of research experience on the politics of water resources management, particularly in developing country contexts including the Mekong region. Key themes of research include water governance, transboundary water conflict and cooperation and water security. She has multi-disciplinary training in international relations, international studies and human geography and is keenly interested in the communication between science and policy.

Bui Lien Phuong is Research Program Coordinator in the Center for Water Resources Conservation and Development, Vietnam. She is responsible for planning and managing projects that aim at promoting river basin management. In addition, she works on research projects related to water governance, climate change and indigenous knowledge. Through these projects, Bui Lien Phuong has also gained experience in capacity building for local people, especially ethnic groups, and in advocacy work with authorities at district, provincial and national levels. She holds a master of environmental sciences from Wageningen University, the Netherlands.

Jakkrit Sangkhamanee is Assistant Professor in the Department of Sociology and Anthropology and the Deputy-Director of MA program in international development studies at the Faculty of Political Science, Chulalongkorn University, Thailand. He has a keen interest in various social and political aspects related to Mekong development. His researches cover a wide range of issues such as transborder development, natural resource management and conflicts, community environmental infrastructure and the sociology of Mekong hydrological knowledge. Jakkrit earned his PhD in anthropology from the Australian National University. His doctoral research dealt with the the politics of knowledge production on water management in northeastern Thailand and the Mekong region.

Thea Sok is Research Associate in the Department of Environmental Science at the Royal University of Phnom Penh, Cambodia. He is also a socio-economic and resettlement specialist at the Environment and Assessment (E&A) Consultant Company. From 2004 to 2011, Thea Sok has worked as a researcher and trainer in the Research and Development Unit at the Cambodian Centre for Study and Development in Agriculture (CEDAC). He obtained a master's degree in rural development management at Khon Kaen University of Thailand in late 2007 and another master's degree in international studies at Tsukuba University, Japan in August 2012. His research interests include international relations, public policy, agriculture and rural development, improved cookstove and wood energy, saving groups, climate change, environment issues, hydropower development and compensation and resettlement.

Niki West has held positions with NGOs, a consulting firm, and a university research centre, and has worked in Canada, the United States, India, Peru and Laos. Niki obtained a Bachelor of Commerce from the University of Alberta, Canada, and a Master of Environmental Management, with a Certificate in International Development Policy, from Duke University, USA. Her research has focused on the water-energy nexus, ecosystem services and corporate social responsibility.

Zha Daojiong is Professor in International Political Economy at the School of International Studies, Peking University (PKU), Beijing, China. He leads the PKU team in the Challenge Program on Water and Food's 'Improving Hydropower Decision-Making Processes in the Mekong' project. Zha Daojiong works closely with Chinese and international colleagues in academic, corporate, government and international agencies on improving political and societal risk management for Chinese companies investing abroad, especially in the extractive resource industry. His research focus is on the politics and policy of China's interaction with the rest of the world, concentrating on energy, food and water. He publishes extensively, studied and taught in the United States and Japan, and has held research positions abroad, including Australia, Singapore and Hong Kong.

Acknowledgements

This volume is an output from a two-year CGIAR Challenge Program on Water and Food project titled 'Improving Hydropower Decision-Making Processes in the Mekong'. The project involved partners from across the Mekong Basin, and was funded by a generous grant from Australian Aid. The editors thank all the writers for their efforts in bringing this book together. We also thank the kindness of many reviewers who provided invaluable comments throughout the project and the development of this book. A special thanks goes to Clare Sanford for her copy-editing, Alan Boatman for the map in the introductory chapter and Tim Hardwick and Ashley Wright from Earthscan for their patience and support.

Abbreviations

ADB	Asian Development Bank
ASEAN	Association of Southeast Asian Nations
ASEM	Asia-Europe Meeting
BIT	bilateral investment treaties
BOO	build-operate-own
BOOT	buy-own-operate-transfer
BOT	build operate transfer
CA	concession agreement
CDM	Clean Development Mechanism
CDRI	Cambodia Development Resource Institute
CGIAR	Consultative Group on International Agricultural Research
CHG	China Huaneng Group
CPP	Cambodian Peoples' Party
CPWF	Challenge Program on Water and Food
CSEZB	Cambodia Special Economic Zone Board
CSO	civil society organizations
CUSRI	Chulalongkorn University Social Research Institute
EAC	Electricity Authority of Cambodia
ECAFE	Economic Commission for Asia and the Far East
ECNEQA	Enhancement and Conservation of National Environmental Quality Act
EDC	Electricité du Cambodge
EDF	Electricité de France
EdL	Electricité du Laos
EGAT	Electricity Generating Authority of Thailand
EIA	environmental impact assessment
EMP	environmental management program
ERC	Energy Regulatory Commission
ESIA	environmental and social impact assessment
ESMAP	Energy sector management assistance program
EVN	Vietnam Electricity
EVNI	Vietnam Electricity International
FDI	foreign direct investment

FIVAS	Association for International Agricultural Research
GDP	gross domestic product
GHG	greenhouse gass
GMS	Greater Mekong Subregion
GoL	Government of Laos
HFO	heavy fuel oil
HHPC	Houay Ho Power Company
HVDC	high voltage direct current
ICEM	International Center for Environmental Management
IDB	international development banks
IEA	International Energy Agency
IMC	Interim Mekong Committee
INGO	international non-governmental organizations
IPP	independent power producers
IRC	Inter-Ministerial Resettlement Committee
IUCN	International Union for the Conservation of Nature
KCC	Key Consultants Cambodia
LHSE	Lao Holding State Enterprise
LMR	Lower Mekong region
LPRP	Lao People's Revolutionary Party
MDG	Millennium Development Goals
MEF	Ministry of Economy and Finance
MIGA	Multilateral Investment Guarantee Agency
MIME	Ministry of Industry Mines and Energy
MoEP	Ministry of Environment Protection
MOIT	Ministry of Industry and Trade
MoLR	Ministry of Land Resources
MONRE	Ministry of Natural Resources and the Environment
MoPI	Ministry of Power Industry
MoU	Memorandum of Understanding
MoWR	Ministry of Water Resources
MoWRPI	Ministry of Water Resources and the Power Industry
MRC	Mekong River Commission
MWRAS	Mekong Water Resources Assistance Strategy
NA	National Assembly
NEPC	National Energy Policy Council
NESDB	National Economic and Social Development Board
NGO	non-governmental organizations
NHI	Natural Heritage Institute
NORAD	Norwegian Agency for Development
NT2	Nam Theun 2 Dam
NTFP	non-timber forest products
NYC	New York City
OCED	Organization for Economic Cooperation and Development
PDA	project development agreement

PNPCA	Procedures for Notification, Prior Consultation and Agreement
PPC	provincial people's committee
PPP	public–private partnerships
PRC	People's Republic of China
REE	rural electricity enterprise
RGC	Royal Government of Cambodia
RMB	renminbi (Chinese currency)
RUPP	Royal University of Phnom Penh
SEA	strategic environmental assessment
SEARIN	Southeast Asia Rivers Network
SIA	social impact assessments
SMEC	Snowy Mountains Engineering Corporation
SOE	state-owned enterprises
SPCC	State Power Corporation of China
SRP	Sam Rainsy Party
STEA	Science, Technology and Environment Administration
TDRI	Thailand Development Research Institute
TGP	Three Gorges Project
THPC	Theun-Hinboun Power Company
UNDP	United Nations Development Programme
UNTAC	United Nations Transitional Authority in Cambodia
VRN	Vietnam Rivers Network
VUSTA	Vietnam Union of Science and Technologies Associations
WB	World Bank
WCD	World Commission on Dams
WCED	World Commission on Environment and Development
WWF	World Wildlife Fund
YCAR	York Center for Asian Research
YPG	Yunnan Provincial Government

1   On dams, demons and development

The political intrigues of hydropower development in the Mekong

Nathanial Matthews and Kim Geheb

> The hazard of building concrete and steel structures that meet technical requirements and fall short in their promotion of a stable life in the basin is great.
>
> (White *et al.* (1962) on Mekong Basin hydropower development)

Introduction

Known as 'the rice basket of Southeast Asia', the Mekong Basin is home to a population of 70 million with 90 distinct ethnic groups (Galipeau *et al.*, 2012). With its extensive wetlands and floodplains, the basin supports the largest inland fisheries in the world with an annual catch of 2.6 million tonnes and over 500,000 tonnes of other aquatic animals (e.g. aquatic insects, amphibians, and molluscs) valued annually at between $3.9 and $7 billion (Hortle, 2007). Over two-thirds of the basin's population are involved in fishing for their livelihoods or to support food security (MRC, 2003). In the Lower Mekong Basin (i.e. those basin areas outside of China), aquatic resources make up between 47 and 80 per cent of animal protein in rural diets (Baran and Ratner, 2007).

The basin has a high level of biodiversity with an estimated 1,200 species of fish and a number of endemic and endangered species. To the economies of southeast Asia, the basin is a source of wealth and power that are underpinned by the hydropower, transport and irrigation associated with the river and its tributaries. Many countries in the basin are also undergoing rapid economic growth and modernization. Hydropower is seen as a key component of this change.

This book provides a political and economic perspective on hydropower development and hydropower decision-making across the Mekong Basin. Drawing primarily from locally based authors using extensive field interviews, document analysis, workshops and surveys, our aim is to shine new light on the political economy and ecology of hydropower development and decision-making in the Mekong. The book is a product of a two-year CGIAR Challenge Program on Water and Food (CPWF) project titled 'Improving Hydropower Decision-Making Processes in the Mekong', funded by Australian Aid. The project's focus was to understand and improve the ways in which decisions were made with respect to hydropower location, coordination, development and operation in the

Mekong Basin. The project included teams from Chulalongkorn University in Thailand, Peking and Yunnan universities in China, the National University of Laos, the Royal University of Phnom Penh in Cambodia, and the Water Recourses Conservation and Development Centre in Vietnam.

This chapter is divided into three sections. First, we present an overview of the historical development of the basin with a focus on hydropower and the key regional organizations. Next, through a political and economic snapshot of the countries within the basin, we add some additional context and background to the arguments presented in subsequent chapters and finally we conclude by summarizing of each of the book's chapters.

An historical survey of Mekong Basin hydropower development and drivers of decision-making

Pre-development era of the river

The 1856 Treaty of Friendship, Commerce and Navigation and the 1893 Treaty for Regulating the Position of the Kingdom of Cambodia were among the first foreign bilateral agreements in the Mekong Basin. These agreements, and subsequent treaties in 1926, 1937 and 1950 focused on the role of navigation and established the thalweg[1] as the precise border between Thailand and Laos.

During this period, the French colonial authority stressed the importance of freedom of navigation to keep open trade routes, mainly for timber. It was also hoped that these river trade routes would eventually provide access to China's wealth and provide a buffer to British colonial expansion in Burma (Osborne, 2006).

In 1950, France, Cambodia, Laos and Thailand signed an agreement to use the waters flowing in their territory for hydropower and irrigation, on condition that these interventions did not impact the 'legitimate interests' of the other countries, or navigation. As Salman (2007) states, the focus on navigational uses of watercourses was prominent during the nineteenth century as many nations were concerned with trade in raw materials. In developing countries, the lack of infrastructure meant that rivers were sometimes the only way to move goods and people quickly across large land areas. With the outbreak of the Second World War in the region in the 1940s and increasing demands for energy to process materials and supply electricity to growing cities, non-navigational uses of watercourses began to take on new importance (Salman, 2007). It was during this time that the Mekong River's hydropower potential started to be recognized.

Early development of the basin and the geopolitics surrounding the formation of the Key River Basin Organization – the 1950s to the 1990s

The Mekong has a long history of development and dialogue among the lower riparian countries dating back over half a century. The first step came in 1957 as part of an economic development drive following the end of the first

Indochina war. The United Nations regional office in Bangkok, the United Nations Economic Commission for Asia and the Far East (ECAFE) studied the basin's hydroelectric and irrigation potential and emphasized the need for cooperative development, including the establishment of a joint body for exchanging information and development plans between the riparian states. ECAFE employed the services of the US Bureau of Reclamation in 1955 to study potential hydropower sites in hopes that the Mekong could be developed into an Asian Tennessee Valley Authority (Osborne, 2006).

Three key studies guided hydropower development at this time. First, in 1957, the UN Economic Commission for Asia and the Far East (ECAFE) released its report on the *Development of Water Resources in the Lower Mekong*. The report recommended the development of the basin's resources through the construction of a number of multi-purpose dams to supply irrigation, generate hydropower, and control flooding. The report reflected the dominant view of the time that dams were modern and progressive (Osborne, 2006).

In 1957, as a result of this initial ECAFE study, Thailand, Laos, Cambodia, Vietnam and an observer from the United Nations Development Programme (UNDP) signed an agreement establishing the Committee for the Coordination of Investigations of the Lower Mekong Basin (Mekong Committee). China was not invited because it was not a member of the United Nations at the time, and Burma was preoccupied with internal politics. From the outset, the Mekong Committee's task was to 'promote, coordinate, supervise, and control water resource development projects in the Lower Mekong Basin' (Nguyen, 1999, 156). The Mekong Committee was the UN's first involvement in the planning and development of an international river basin (Jacobs, 1995).

Soon after the UN became involved in the basin, it conducted its own study on the development of the Mekong, the *Programme of Studies and Investigation for Comprehensive Development of the Lower Mekong River Basin* (United Nations, 1958). This study, also known as the Wheeler Report, was headed by Raymond Wheeler of the US Army Corps of Engineers. The Army Corps were active in much of the United States' dam development in the early twentieth century (Jacobs, 1995). This report reaffirmed the Mekong's huge hydropower potential and outlined an ambitious plan to develop both mainstream and tributary dams across the basin. A precursor to this plan, however, was a five-year data collection programme to guide project planning. To some degree, the report conflicted with the ECAFE report. While both reports supported economic development including hydropower dams, the ECAFE report was focused more on tributary dams and smaller-scale projects, while the 1958 study was more cautionary and was informed by existing, detailed studies. As Nguyen (1999, 57) states: 'For some, the Wheeler Report's goals appeared much too ambitious and beyond the member countries' capacity.'

The third and final report that guided development at this time was a Ford Foundation-sponsored report by Gilbert White (White *et al.*, 1962), *Economics and Social Aspects of Lower Mekong Development*. The White Report went beyond the engineering and technical considerations of the previous two reports to look at the potential environmental and social impacts of development. For White (1962), the report was a way of illustrating Asia's first large-scale efforts to study the economic, institutional and social aspects of development prior to development occurring.

From these reports, the Mekong Committee developed an agenda that promoted irrigation, hydropower and flood control projects as a form of poverty alleviation. During the 1960s, the United States was heavily involved in the Mekong Committee funding as the largest non-riparian donor, with Thailand as the largest regional donor. The United States' interest in funding the Mekong Committee and its development projects stemmed primarily from its desire to curb communist influence in the region (Chi, 1997; Hirsch, 2006).

Hydropower development on the Mekong progressed slowly. The Mekong Committee completed a small number of minor tributary irrigation, multi-purpose and hydropower dams in Laos and Thailand during the 1960s and early 1970s, but many projects were inhibited by a lack of finance within the region, a lack of detailed economic and environmental baseline studies as well as lack of skilled technical workers (Jacobs, 1995). The Mekong Committee also began flood forecasting and basin-wide data collection programmes.

The stage was set for hydropower to begin in earnest from the 1970s. At this time, the Mekong Committee drafted the Indicative Basin Plan, the first Basin Development Plan for the region. This plan recommended 180 possible projects, including four mainstream dams. Political tensions, however, in the 1970s, including conflict in Laos, Cambodia and Vietnam, reduced the Committee's funding and once again limited project development.

In 1975, the Mekong Committee ratified the Joint Declaration of Principles that further defined the principles, norms, rules and decision-making procedures of development. The Joint Declaration stated that mainstream waters could not be developed without prior approval of the other basin states through the Committee; this gave each member a veto power over mainstream dam development (Mekong Committee, 1975).

In 1976 and 1977 Laotian, Cambodian, and Vietnamese representatives were absent from the Committee due to a shift in regional governments and disruptions associated with armed regional conflicts. Although Laos and Vietnam rejoined in 1978, Cambodia did not. Cambodia's split from the Committee meant that mainstream dam development could not legally proceed because of the Joint Declaration required approval from all members before mainstream waters were dammed.

From 1978 to 1994, the Committee re-established itself as the Interim Mekong Committee (IMC), consisting of Thailand, Laos and Vietnam. Moving away from basin-wide development, the IMC focused on data collection and domestic projects in the remaining three member states (Jacobs, 1995).

In 1995, after Cambodia rejoined, a lengthy negotiation process was initiated via the Mekong River Commission (MRC). The MRC was established with the Mekong Agreement which, once again, re-engaged the four lower riparian states. During this time, China and Burma were invited to join the MRC, but they declined, becoming 'dialogue partners' in 1996 instead. The MRC's mandate was 'to cooperate in all fields of sustainable development, utilization, management and conservation of the water and related resources of the Mekong River Basin', and 'to ensure reasonable and equitable use' of the Mekong River system (MRC, 1995).

The Mekong Agreement and the Mekong River Commission

The lengthy negotiation leading up to the establishment of the MRC saw important changes in its capacity to manage mainstream dam development. The early 1990s marked the rapid growth and industrialization of Thailand's economy. This rapid growth accompanied pressure on Thailand to accelerate the development of its hydroelectric power and irrigation resources. These developments, which included plans for Mekong mainstream diversions and dams, were not well received by Thailand's neighbours.

One such controversial project was the Khong Chi Mun scheme, a large inter-basin water diversion project in Thailand's northeast that proposed to divert large volumes of water from the Mekong, the Chi and Mun river systems through canals to 'turn the Northeast green' (Sneddon, 2003). Vietnam and Cambodia were concerned about the project and Thailand's ambitions to develop the mainstream. They contended that diverting the Mekong would be harmful to the Tonle Sap and the Mekong Delta, the breadbasket of Vietnam (Makim, 2002). The conflict of interest between Thailand, Cambodia and Vietnam contributed to a four-year deadlock in the negotiations leading to the 1995 Agreement. Thailand's first move was to link Cambodia's re-admission to the process to the renegotiation of the rules and laws governing the MRC (Posey, 2005). Thailand held a powerful position in this negotiation as an upper riparian with a strong economy. Vietnam and Cambodia were both downstream with important economic and food security (fishing and agricultural) interests to protect. Thailand pushed to remove the UNDP as the executive agent of the body. This helped to neutralize the leadership of the United States and the international community represented by the UN in the Committee (Makim, 2002). Finally, Thailand demanded that the new Mekong Agreement dissolve the existing veto power that each nation had over mainstream dam development.

This was initially opposed by a coalition of the three downstream riparian states, Cambodia, Laos and Vietnam. Thai bilateral diplomacy, however, eventually won over Cambodia and Laos, and Vietnam was forced to accede to the removal of the veto and its replacement with a 'prior notification' of development.

With the 1995 Agreement, the MRC's institutional structure developed into three permanent bodies: the Council, the Joint Committee and the Secretariat. As a result of the new protocols, MRC member countries had to now notify each other if they wished to engage in any major infrastructure developments (such as hydropower schemes) on the Mekong or tributaries, particularly if those developments might have significant transboundary impact for people or the environment downstream. These new protocols are called the Procedures for Notification, Prior Consultation and Agreement (PNPCA) (MRC, 2003).

The PNPCA protocols state that if a member country is to build hydropower dams on a Mekong tributary, it must first notify the Joint Committee of the MRC. Any mainstream development, such as the currently proposed 11 mainstream hydropower dams, are subject to notification and a consultation agreement, with the aim of arriving at an agreement by the Joint Committee of the MRC. 'National Mekong Committees' are the submitting parties. The PNPCA is triggered when the preparation of a mainstream dam advances to the stage where the member country makes a submission to the MRC. In considering proposals for mainstream hydropower developments, the Joint Committee must avoid inter-state disputes by resolving and determining if the development (MRC, 2003):

- optimizes water use;
- provides better benefits than can be derived through cooperation and trade-offs;
- has an established right of claim against further proposed uses;
- assesses the potential impacts on multi-stakeholder's rights and interests;
- provides for planning security.

During the wet and dry seasons, specific notification requirements are in place on the mainstream of the Mekong River. During the wet season, intra-basin uses are subject to notification to the Joint Committee, and inter-basin diversions are subject to prior consultation with the aim of arriving at an agreement by the Joint Committee. Conversely, during the dry season, intra-basin uses are subject to prior consultation with the aim of arriving at an agreement by the Joint Committee. Furthermore, the Joint Committee, prior to any proposed diversion, shall agree upon any inter-basin diversion project. Should there be a surplus of water available in excess of the proposed uses by all member states in any dry season (verified and unanimously confirmed as such by the Joint Committee), an inter-basin diversion of such surplus can be made, subject to prior consultation (Burchi and Spreij, 2003).

Although the PNPCA has removed the power of any member state to veto mainstream dams or diversions of another, the 1995 Mekong Agreement stresses in the statute's definition of terms that the removal of the veto does not amount to a 'unilateral right to use water by any riparian without taking into account other riparian state rights' (MRC, 1995).

Throughout the early 2000s and up to 2014, the MRC has continued to focus on monitoring and data collection activities, and fisheries, transportation and flood control programmes to develop regional expertise. The MRC has also expanded its mandate to focus on Integrated Water Resources Management (an integrated basin development approach) and, as discussed in Chapter 6, as a controversial mediator of mainstream dam development. While the MRC is the basin's largest managerial agency, there are other basin-wide initiatives with a focus on hydropower development.

The Greater Mekong Subregion Programme

Initiated in 1992, the keystone programme of the Asian Development Bank (ADB) in the Mekong Basin is the Greater Mekong Subregion (GMS) Programme. The GMS Programme was created by Cambodia, Laos, Burma, Thailand, Vietnam and Yunnan Province of the People's Republic of China with assistance from the ADB. The GMS Programme is a private sector-led economic cooperation programme with a long-term goal 'to promote development through closer economic linkages' (ADB, 2009).

The nine priority areas of GMS interconnectivity are transport, tele-communications, energy, tourism, human resources, development, environment, agriculture, trade and investment (ADB, 2012). Projects completed between 1992 and 2011 have totalled costs of approximately $10 billion, of which about $5 billion has been put forward by the ADB, with the balance comprising regional and international investment (ADB, 2012). In terms of hydropower development, the chief initiatives of the GMS Programme have been the Mekong Power Grid and associated support for the Nam Theun 2 Dam in Laos. The Mekong Power Grid is part of the Regional Power Interconnection and Power Trade Arrangements of the GMS Programme. The objective of the grid is to 'promote a commercially-based energy system that reliably and competitively supplies electricity to all areas of the Subregion in a manner that minimizes environmental and social costs' (ADB, 2011). The grid consists of a series of transmission lines and hydropower schemes to allow Thailand and Vietnam to purchase electricity from Yunnan Province, Laos, Burma and Cambodia. The ADB has estimated the cost of the grid at $43 billion.

The Nam Theun 2 (NT2) dam is partially funded by the ADB's programme and is a component of the Mekong Power Grid. The development of the NT2, which was commissioned in 2010, paved the way for private-sector financing and construction of hydropower in Laos by establishing procedures, contracts and tariffs for large-scale investment in the country (see Chapter 7). The GMS Programme is part of a broader set of regional initiatives to bring the diverse nations of the region together to synergize mutual economic growth. The Association of Southeast Asian Nations, modelled after the European Union, expands beyond the borders of the Mekong Basin to attempt to foster collaboration within southeast Asia.

The Association of Southeast Asian Nations

The Association of Southeast Asian Nations (ASEAN) was established in 1967 between Indonesia, Malaysia, Philippines, Singapore and Thailand. By 1999, ASEAN had expanded to include Brunei, Laos, Burma, Cambodia and Vietnam, thus encompassing a total of ten countries. The aims of ASEAN can be summarized as the promotion of economic growth, peace and stability across the region. The principles governing the organization are non-interference in the internal affairs of member states and respect for states' sovereignty and independence (Tong and Chong, 2010). ASEAN contributes significantly to economic growth in the Mekong Region. In 2010, the China–ASEAN Free Trade Agreement formed the world's largest free trade area, comprising 1.9 billion consumers and US $4.3 trillion in trade (Tong and Chong, 2010).

In terms of hydropower in the Mekong Basin member countries, ASEAN promotes a renewable energy target of 15 per cent among its members, of which hydropower is a key component and hydropower issues, including the Xayaburi Dam, have been on the agenda at regional meetings (Abdullah, 2005). ASEAN has also been a supporter of the ADB's GMS Programme and the Mekong Power Grid.

Current state of hydropower development in the basin

As outlined above, hydropower in the Mekong has been under exploration and debate for over half a century, but various combinations of insufficient capital, weak technical capacity, conflict and political instability and environmental concerns have impeded its expansion. This section offers a brief overview of some of the key issues in hydropower development in the Mekong Basin. In 1995, following the ratification of the 1995 Mekong Agreement and the reformation of the MRC, it appeared that both mainstream and tributary dam development would gather pace. Shortly thereafter, development plans were once again put on hold due to the 1997–8 Asian financial crisis that precipitated a serious devaluation of southeast Asian currencies and eventually led to a global economic slowdown.

With the gradual recovery from the Asian financial crisis, hydropower appeared on the development agenda once again. In 2000, however, the influential World Commission on Dams (WCD) report (2000) called for a rethink about the risks and costs of large-scale dams across the world, making hydropower an unpopular target for funding. Adding to the financing issues, from the mid-1990s to the early 2000s, the World Bank (which funded the WCD Report) ceased funding large-scale hydropower projects due to environmental and social criticisms levied by the NGO sector (Park, 2005).

From the mid-2000s, a number of factors contributed to a reawakening of the hydropower agenda both globally and in the Mekong Basin. The hydropower resurgence emerged in part due to a global push for clean, renewable energy sources, increased availability of capital, potential profits, and rising electricity demands.

In 2005, the World Bank returned to large-scale infrastructure development in the region with its involvement in the Nam Theun 2 hydropower project in Laos. In a 2006 influential policy report, the World Bank called for regional hydropower to be expanded and estimated that only 10 per cent of the Mekong Basin's hydropower potential was in use (WB/ADB, 2006). These estimates further state that the basin has the 'flexibility and tolerance' to handle an increase in hydropower development including mainstream dams (WB/ADB, 2006).

In 2005, there were an estimated 25 hydropower projects completed on the Mekong tributaries, with approximately 35 in planning or feasibility stages, with two mainstream dams complete on the upper Mekong River in China – a relatively small number considering the size of the basin. By 2014, at least 28 more dams were under construction, five mainstream dams were completed in China, the lower basin's first mainstream dam, the Xayaburi, was under construction and a further 150 projects were identified as probable development sites, including an additional ten mainstream dams on the lower Mekong (see Figure 1.1) (King *et al.*, 2007; Johnston and Kummu, 2012; CPWF, 2014).

Diversity of the six Mekong riparians and their position in the basin

China

Led by a single-party communist government, and with a population of 1.4 billion, China is a political and economic powerhouse. In 2012, China's gross domestic product (GDP) was $8227 trillion with a growth rate of 7.8 per cent (World Bank, 2014). The importance of the Mekong in China is within Yunnan Province where the Mekong provides electricity through dozens of hydropower dams as well as water for irrigation. Approximately 10 million people live within the Mekong Basin in Yunnan Province. In China, where the river and basin are born, the Mekong River is called the Lancang Jiang. As shown in Figure 1.1, the river is narrow as it passes through China, dropping over 4,500 metres through deep gorges before beginning to expand and slow down once it has passed the southern Yunnan Province border.

Although almost half of the Mekong River's length passes through China, however, the contribution to flow is only 16 per cent. This small percentage is important because it represents 34 per cent of the flow during the dry season (Nesbitt et al, 2004). As the Mekong flows south and the basin begins to take shape, it skirts the border with Myanmar, gaining 2 per cent of its flow before entering into Laos where the mountainous landscape and dozens of tributaries swell its volume (Nesbitt *et al.*, 2004).

Laos

Laos is a single-party communist state with a population of 6.45 million. Laos's GDP is $9.41 billion and it has a growth rate of 8.4 per cent (World Bank, 2014). Approximately 80 per cent of the population is rural and 27.6 per cent exist below the level defined by the World Bank as living in poverty (World Bank, 2014). The Mekong and its tributaries are the lifeblood of Laos, with approximately five million people living within the basin. In fact, Laos is situated almost entirely within the Mekong Basin, with the river gaining 35 per cent of its flow within Laos' borders (Hirsch, 1995). Laos has a high population of ethnic minorities who live close to the mainstream or the Mekong tributaries, relying primarily on wild fish and subsistence agriculture for their food security (Nguyen Khoa *et al.*, 2005). For example, 71 per cent of rural households rely on

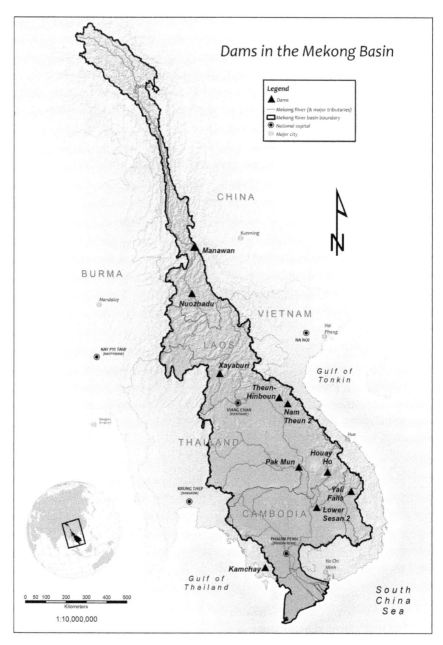

Figure 1.1 Case study dams: the main dams analysed in the book

Source: Geo-Sys (Lao) Company Ltd.

fisheries for either subsistence or as a source of income, while fish provided by the basin are second only to rice for food security and income (Nguyen Khoa *et al.*, 2005). As the Mekong River travels south it forms the border between much of western Laos and Thailand.

Thailand

Thailand is a democratic upper-middle income country with a population of 66.79 million. The Thai GDP ws $366 billion and the country had a growth rate of 6.5 per cent in 2012 (World Bank, 2014). Thailand contributes 18 per cent of the Mekong's flow, predominantly with waters emanating from the mountainous north of Thailand. In the northeast of Thailand, where the basin covers a significant tract of farmland, the semi-arid climate and intensive agriculture of the Mun and Chi Basins drain 15 per cent of the Mekong Basin (Hirsch, 1995). Approximately 25 million people live within the Mekong Basin in Thailand. Agriculture, primarily rice, and fishing dominate this segment of its rural economy. The Mekong then continues its journey south into Cambodia.

Cambodia

Cambodia is a low-income country with a population of 14.85 million. The Cambodian GDP is $14.04 billion and the country had a growth rate of 7.3 per cent in 2012. The government is a constitutional monarchy that operates as a parliamentary representative democracy, but the prime minster of Cambodia, Hun Sen, has held office since 1985. In Cambodia, 86 per cent of the country is located in the basin and this area is home to approximately 10 million people.

In Cambodia, the river landscape is mainly flat and the river forms into a delta as it moves past Phnom Penh. Cambodia contributes 18 per cent flow to the Mekong River. The great Cambodian lake, Tonle Sap, is fed by the Mekong (Kummu and Sarkkula, 2008). Tonle Sap grows sevenfold from 2,500 square kilometres and a depth of 0.5 metres to 16,000 square kilometres and depth of 7 to 10 metres during the monsoon (Kummu and Sarkkula, 2008). The lake supplies 15 to 20 per cent of the basin's fish catch and is an important feeding ground for fish (Lamberts, 2006). The Cambodian population is heavily reliant on fish for food security, with 80 per cent of the population's animal protein deriving from wild fish (Hortle, 2007). Finally, the river enters Vietnam, through the Mekong Delta.

Vietnam

Vietnam is the furthest downstream riparian. It is a one-party communist state with a population of 88.7 million and its economic growth rate was 5.2 per cent in 2012 (World Bank, 2014). The country covers an area of 331,210 km^2 and contributes 11 per cent to the flow of the Mekong River. The importance of the Mekong Basin and its water resources to Vietnam's food and economic security cannot be underestimated (Käkönen, 2008). The Mekong Delta contains 37 per cent of Vietnam's cultivated land area, producing more than 50 per cent of Vietnam's fishery, 60 per cent of the country's fruit and over 30 million tonnes of agricultural products (Käkönen, 2008). The Delta is home to a population of approximately 20 million.

Book structure

Chapter 2, 'Framing a political ecology of Mekong Basin hydropower development', begins by outlining the guiding framework for the book, which also underpinned the project on which this work is based. Political ecology was chosen because its analysis of power relations between different scales vis-à-vis the environment and its development, and allows for a cumulative understanding of the impacts and benefits of decisions made with respect to hydropower development. Incorporating a strong ecological component to the guiding framework of the approach was deemed essential in the Mekong, because millions of people's lives are deeply intertwined with the environment. Chapter 2 demonstrates the utility of the political ecology approach through a detailed literature review that sets the context for analysis in subsequent chapters.

Chapter 3, 'A political ecology of hydropower development in China', begins by analysing Chinese justifications for hydropower in China in terms of science, energy security, statistics and debates. The first half of the chapter offers a political ecology analysis of hydropower growth and the implications of the lessons learned from domestic hydropower development across China. The second half of the chapter examines China's hydropower development beyond its borders in the Lower Mekong region, addressing scope and controversy, domestic push factors, ideational self-justifications and efforts at internationalization.

Chapter 4, 'From Manwan to Nuozhadu: the political ecology of hydropower on China's Lancang Jiang', examines the narratives and power relations driving hydropower development on the upper Mekong River. The first part of this chapter traces the trajectory of hydropower from the Manwan Dam, the first mainstream on the Mekong, to the Nuozhadu Dam, which began operation in 2013. The chapter highlights the changing social, economic and political contexts for China's growth. The second part discusses the social, economic and policy settings that significantly alter hydropower sector development and how negative social and environment impacts produce policy changes. The chapter details how economic reform has resulted in national state-owned companies gaining power over the local government, at the same time allowing national agencies to gain control over China's development agenda.

Chapter 5, 'From Pak Mun to Xayaburi: the backwater and spillover of Thailand's hydropower politics', analyses the changing dynamics and the complexities of the actors, regulation and cooperation, technology and narratives that have played, and still play, a crucial role in shaping the decision-making process in hydropower enterprises in Thailand. Employing the two analogies of 'backwater' and 'spillover' to understand the ecology of hydropower politics, the chapter investigates and compares two study cases, the Pak Mun and Xayaburi dams. It argues that very little changed during the two decades between the Pak Mun Dam and the Xayaburi. Much of the politics, science, actors and decision-making processes remain the same. The chapter concludes by arguing that the emerging challenge of hydropower decision-making improvement lies in the shift towards the transboundary nature of hydropower development within the Mekong region.

Chapter 6, 'The invisible dam: hydropower and its narration in the Lao People's Democratic Republic', uses the political ecology approach to critically analyse the narratives behind the polarized debate surrounding hydropower development in Laos. Drawing from policy and media statements from the government of Laos, NGOs and civil society, the chapter shines light on the drivers behind the increasingly divergent and fantastic claims regarding the impact of mainstream dams on the people and environment of the Mekong Basin.

Chapter 7, 'Whose risky business? Public–private partnerships, build-operate-transfer and large hydropower dams in the Mekong region', adopts a political economy approach to offer insights into how private sector investments have shaped hydropower development and its impacts in the Mekong Basin. The chapter uses case studies from both Laos and Cambodia to illustrate how the build-operate-transfer (BOT) and independent power producers (IPP) models have empowered certain actors over others.

Chapter 8, 'The politics of the Lower Sesan 2 Dam in Cambodia', presents a political ecological overview of hydropower development in Cambodia with a specific focus on the Lower Sesan 2 Dam. The chapter analyses key stakeholder roles, narratives and perspectives surrounding the emergence and spread of hydropower across Cambodia. The chapter further analyses protest movements and the impediments to resettlement in Cambodia when animism and culture are considered.

Chapter 9, 'Rethinking development narratives on hydropower in Vietnam', goes beyond the current academic literature examining the hydropower resettlement of the Central Highlands to focus on the broader issues of shifting politics between the state and private sector surrounding dam development in Vietnam. The chapter begins by giving an overview of the role of the Central Highlands in Vietnam before and after the country's reunification in 1975. It then explores the narratives, policy changes and actors surrounding hydropower in Vietnam in general and the Central Highlands in particular. The chapter concludes by focusing on problems caused by hydropower development in Vietnam, including a discussion on deforestation, resettlement, social and cultural destruction and unequal distribution of costs and benefits with a specific focus on the Sesan River.

Conclusion

Over 60 per cent of the planet's freshwater resides and flows within transboundary rivers and aquifer systems, and approximately 40 per cent of the world's population lives within these basins. Within the 263 transboundary basins around the world, approximately two-thirds span developing economies. It is in the developing economies of Asia, Africa and Latin America, along transboundary rivers, where most of the remaining hydropower development potential exists. According to a 2011 International Energy Agency (IEA) report, global hydropower could grow as much as 85 per cent by 2050, an increase of 150 to 200 GW of new generating capacity (IEA, 2011).

Hydropower is a potentially cheap source of electricity, and is often argued to have lower greenhouse gas emissions compared to burning hydrocarbons (Barros *et al.*, 2011). Dams can also, however, have significant negative impacts. They can displace populations, destroy cultures, and alter the flow, temperature, water quality, sediment loads and ecosystems that depend on water, rendering it potentially unusable for irrigation, environmental services, fisheries and livelihoods (WCD, 2000).

We hope that this book will help a variety of actors better understand the political and economic drivers behind hydropower development and decision-making and its diverse impacts on people and the environment in the Mekong Basin. We believe that improved understanding is an important step in helping actors in the Mekong Basin and further abroad reach more equitable and inclusive water management decisions.

Note

1   A 'thalweg' is a fluvial geomorphology term that denotes the deepest continuous inline within a watercourse or valley.

References

Abdullah, K. (2005) 'Renewable energy conversion and utilization in ASEAN countries', *Energy*, 30(2–4): 119–28.

ADB (Asian Development Bank) (2009) 'The GMS programme', www.adb.org/GMS/Program/default.asp, accessed Aug 2010.

ADB (Asian Development Bank) (2011) 'Greater Mekong Subregion, regional power interconnection and power trade arrangements', www.adb.org/GMS/Projects/flagshipE.asp, accessed April 2011.

ADB (Asian Development Bank) (2012) 'GMS program overview', www.adb.org/countries/gms/overview, accessed April 2012.

Baran, E., and Ratner, B. (2007) *The Don Sahong Dam and Mekong Fisheries: A Policy Brief from the World Fish Center*, Penang: World Fish Center.

Barros, N., Cole, J., Tranvik, L., Prairie, Y., Bastviken, D., Huszar, V., del Giorgio, P., and Roland, F. (2011) 'Carbon emission from hydroelectric reservoirs linked to reservoir age and latitude', *Nature Geoscience*, 4(9): 593–6.

Burchi, S., and Spreij, M. (2003) *Institutions for International Freshwater Management*, Paris: United Nations Educational, Scientific and Cultural Organization.

Chi, B. K. (1997) 'From committee to commission: The evolution of the Mekong River Agreements', PhD thesis, Law Faculty, University of Melbourne.

CPWF (CGIAR Challenge Program on Water and Food) (2014) 'Map of hydropower and reservoirs in the Mekong', https://mekong.waterandfood.org/archives/2648, accessed April 2014.

Galipeau, A., Mark, I., and Bryan, T. (2012) 'Dam-induced displacement and agricultural livelihoods in China's Mekong Basin', unpublished manuscript.

Hirsch, P. (1995) 'Thailand and the new geopolitics of Southeast Asia: Resource and environmental issues', in J. Rigg (ed.), *Counting the Costs: Economic Growth and Environmental Change in Thailand*, Singapore: Institute of Southeast Asian Studies, pp. 235–59.

Hirsch, P. (2006) 'The Mekong River Commission and the question of national interest(s)', *Watershed*, 12(1): 20–5.

Hortle, G. (2007) *Consumption and Yield of Fish and Other Aquatic Animals from the Lower Mekong Basin*, MRC Technical Paper 16, Vientiane: Mekong River Commission.

IEA (International Energy Agency) (2011) *Clean Energy Progress Report*, Washington, DC: US Department of Energy, Energy Information Administration.

Jacobs, J. (1995) 'Mekong Committee history and lessons for river basin development', *Geographical Journal*, 161(2): 135–48.

Johnston, R., and Kummu, M. (2012) 'Water resource models in the Mekong Basin: A review', *Water Resources Management*, 26(2): 429–55.

Käkönen, M. (2008) 'Mekong Delta at the crossroads: More control or adaptation?', *AMBIO*, 37(3): 205–12.

King, P., Bird, J., and Haas, L. (2007) *The Current Status of Environmental Criteria for Hydropower Development in the Mekong Region: A Literature Compilation*, Vientiane: WWF Living Mekong Program.

Kummu, M., and Sarkkula, J. (2008) 'Impact of the Mekong River flow alteration on the Tonle Sap flood pulse', *AMBIO*, 37(3): 185–92.

Lamberts, D. (2006) 'The Tonle Sap Lake as a productive ecosystem', *Water Resources Development*, 22: 481–95.

Makim, A. (2002) 'Resources for security and stability? The politics of regional cooperation on the Mekong, 1957–2001', *Journal of Environment and Development*, 11(1): 5–52.

Mekong Committee (1975) *Joint Declaration of Principles for Utilization of the Waters of the Lower Mekong Basin*, Bangkok: Mekong Committee.

MRC (Mekong River Commission) (1995) *Agreement on the Cooperation for the Sustainable Development of the Mekong River Basin*, Chiang Rai: Mekong River Commission.

MRC (Mekong River Commission) (2003) *State of the Basin Report*, Phnom Penh: MRC.

Nesbitt, H., Johnston, R., and Solieng, M. (2004) 'Mekong river water: Will river flows meet future agriculture needs in the Lower Mekong Basin?', in S. Veng, E. Craswell, S. Fukai and K. Fischer (eds), *Water in agriculture*, ACIAR Proceedings, 116: 86–104, Canberra: Australian Center for International Agricultural Research.

Nguyen, T. D. (1999) *The Mekong River and the Struggle for Indochina: Water, War and Peace*, Westport, CT: Praeger.

Nguyen Khoa, S., Lorenzen, K., Garaway, C., Chamsingh, B., Siebert, D. J., and Randone, M. (2005) 'Impacts of irrigation on fisheries in rain-fed rice-farming landscapes', *Journal of Applied Ecology*, 42: 892–900.

Osborne, M. (2006) *Mekong*, Crows Nest, New South Wales: Allen & Unwin.

Park, S. (2005) 'Norm diffusion within international organizations: A case study of the World Bank', *Journal of International Relations and Development*, 8(2): 111–41.

Posey, D. (2005) 'Defining interests: The Mekong River Commission', *International Policy Solutions*, 2(1): 1–25.

Salman, S. M. A. (2007) 'The Helsinki Rules, the UN Watercourses Convention and the Berlin Rules: Perspectives on international water law', *Water Resources Development*, 23(4): 625–40.

Sneddon, C. (2003) 'Reconfiguring scale and power: The Khong Chi Mun project in northeast Thailand', *Environment and Planning A*, 35: 2229–50.

Tong, S., and Chong, C. (2010) *China-ASEAN Free Trade Area in 2010: A Regional Perspective*, EAI Background Brief, 519, Singapore: National University of Singapore.

United Nations (1958) *Programme of Studies and Investigation for Comprehensive Development of the Lower Mekong River Basin*, TAA/AFE/3, Bangkok: Report of a United Nations Survey Mission.

WB/ADB (World Bank/Asian Development Bank) (2006) *Future Directions for Water Resources Management in the Mekong River Basin: Mekong Water Resources Assistance Strategy*, Manila: Asian Development Bank.

WCD (World Commission on Dams) (2000) *Dams and Development: A New Framework for Decision-Making*, Report of the World Commission on Dams, London: Earthscan.

White, G. F., deVries, E., Dunkerley, H. B., and Krutilla, J. V. (1962) *Economic and Social Aspects of Lower Mekong Development*, Bangkok: Report to the Mekong Committee.

World Bank (2014) *Country Profiles*, www.worldbank.org/en/country, accessed March 2014.

2 Framing a political ecology of Mekong Basin hydropower development

Nathanial Matthews and Kim Geheb

Introduction

This chapter sets out to frame a political ecology of hydropower in the Mekong River Basin. It considers how political ecology is an appropriate approach to analysing hydropower development in the Mekong River Basin. We explore the emergence of political ecology theory and how it treats narratives and scale, before analysing its application to water resource allocation. Specifically, we will examine how the ecological elements in resource allocation are in themselves a political process.

Political ecology initially emerged in the field of geography and has since developed into an interdisciplinary and diverse approach that draws from evolving ideas in geography, anthropology, human ecology and environmental history. The diversity in political ecology is illustrated by the different understandings of the discipline. For Robbins (2004), Paulson and Gezon (2005) and others, political ecology is a normative approach that focuses on issues such as justice and human rights, tending to be pro-poor and concerned with marginalized groups, and promoting environmental priorities. While Swyngedouw (1999) and Forsyth (2003) among others, eschew any explicitly normative claims and use political ecology solely as a critical approach. In this sense, political ecology has its own political ecologies. At its heart, political ecology seeks to explain the 'complex relations between Nature and Society through careful analysis of social forms of access and control over resources' (Watts and Peet, 2004, 4). As such, central to political ecology is the idea of *access* to natural resources, and how some people, some of the time, are able to control it to exclude others. Political ecology draws from both politics and economic theory to understand who stands to win or lose from environmental change across different scales. This cross-scalar analysis focuses on linking the local to the global with respect to competition and conflict over natural resources. It aims to explain the rationale of actors in political, social and environmental contexts.

Political ecology also analyses power. Power has many definitions and, at least in the literature, many facets. French and Raven's (1959) now classic study focused on the foundations on which power is based; in some cases, the texture of power is all too real – the application of force, and the bringing to bear of arms. But the everyday forms of power that, over the longer term, yield profound outcomes for ecologies and societies are more commonly the stuff of political ecology. Dahl (1957) posits that power is the ability of actor 'A' to get actor 'B' to do what 'A' wants, and this is probably accurate, although the cadence in the definition tends to belie the many subtle and nuanced forms of power and domination.

While political ecology can politicize environmental change, its critical analysis of scale and narratives also helps to illuminate the mechanisms that result in power asymmetries. Like many disciplines, political ecology has struggled with the idea of scale because of its 'relationality' – how people at local scales experience nature, versus how it is experienced by government ministers. The approach is, nevertheless, useful in analysing Mekong Basin hydropower development because hydropower is heavily scaled: international corporations and banks funding and developing hydropower, yielding national level benefits, localized changes to ecologies and community life, and regional cumulative impacts that may be good for some, and detrimental to others.

Much political ecology is also archaeological, thereby introducing an additional element to its analysis: time. It frequently explores the origins of the 'current state of things' over years or decades, thereby charting the history of change and the ebb and flow of conflict and resolution. The Mekong's hydropower potential has been discerned since the 1960s; and the need for humans to exercise their mastery over its nature for much longer (Wong, 2010).

In the Mekong Basin, people and the environment are inextricably linked. Approximately 80 per cent of the basin's population lives in rural communities deriving their animal protein and livelihoods from wild fish and other important aquatic ecosystem services. The politics and economics that shape hydropower development often directly and immediately affect the environment, livelihoods and food security. By drawing attention to the winners and losers in human-induced environmental change processes in the Mekong Basin, political ecology provides a way to critically analyse the power relations at play in the particular economic, political and environmental contexts of the basin. In this sense, political ecology sets out to expose the truth behind hydropower decisions, either because it is simply unknown; or because it is veiled behind narratives concerning development, progress, modernity and improved livelihoods for all.

In hydropower development there are often significant environmental impacts and substantial opportunities for actors to increase their power, either monetarily or politically. Through its analysis of structures that enable and drive the distribution of power and impacts, political ecology enables the identification of insights into the often poorly understood realm of the politics of hydropower development.

Political ecology as an emerging theory

'Political ecology' was first mentioned in Russett (1967, p. vii), who defined it as 'the relation of organisms or groups of organisms to their environment' as he sought to explore 'some of the relations between political systems and their social and physical environment'. Russett's early use of the term did not encompass conservation or the natural environment. Over the years, the definition of political ecology has taken on new meanings. Subsequent analysis of the term occurred through the 1960s and 1970s (see Russett, 1967; Wolf, 1972; Miller, 1978; Cockburn and Ridgeway, 1979). Our current understanding of the approach, however, draws primarily from its application during the 1980s.

In the late 1980s, the term 'political ecology' was employed by scientists studying natural resource management as an approach to move beyond apolitical and neo-Malthusian explanations of environmental change (Watts and Peet, 2004). Conventional wisdom often blamed farmers or local practices and overpopulation for the degradation of resources. These highly simple, cause-and-effect narratives proved extraordinarily powerful. Ideas surrounding the 'tragedy of the commons' (Hardin, 1968) and the 'population bomb' (Ehrlich, 1968) proved immensely influential. Incorporating an approach that brought together politics and ecology allowed researchers to think 'about questions of access and control over resources' and 'how this was indispensable for understanding both the forms and geography of environmental disturbance and degradation, and the prospects for green and sustainable alternatives' (Watts and Peet, 2004, 6).

Blaikie and Brookfield (1987) used a largely structural, empirically oriented, political ecological approach with 'chains of explanation' to identify that soil erosion in African farming villages was driven by broader political economic forces. Blaikie and Brookfield's approach to political ecology relies heavily on ecological perspectives. The chain of explanation offers a method to examine the structures, actors, relationships and asymmetries of power across different scales. The chain of explanation

> starts with the land managers and their direct relations with the land (crop rotations, fuel wood use, stocking densities, capital investment and so on). The next link concerns their relations with each other, other land users, and groups in the wider society who affect them in any way, which in turn determines land management. The state and the world economy constitute the last links of the chain. Clearly then, explanations will be highly conjectural, although relying on theoretical bases drawn from natural and social science.
>
> (Blaikie and Brookfield, 1987, 27)

In Figure 2.1, Blaikie and Brookfield's chain of explanation is applied to a hypothetical hydropower development in the Mekong Basin. Although the chain of explanation draws from conjecture, it offers a method of perceiving relationships between different scales from the ecosystem through to the larger political economy.

By examining how the chain of explanation is adapted to hydropower development, we are able to see the structures, actors and scale involved.

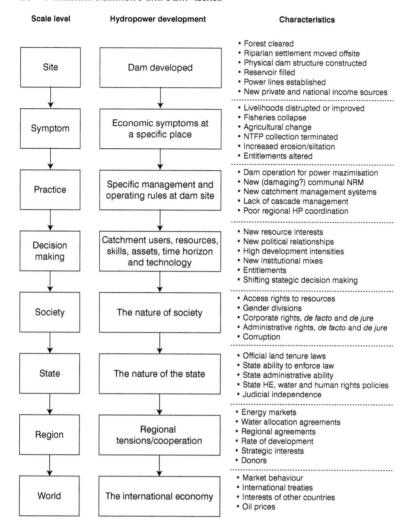

Scale level	Hydropower development	Characteristics
Site	Dam developed	• Forest cleared • Riparian settlement moved offsite • Physical dam structure constructed • Reservoir filled • Power lines established • New private and national income sources
Symptom	Economic symptoms at a specific place	• Livelihoods distrupted or improved • Fisheries collapse • Agricultural change • NTFP collection terminated • Increased erosion/siltation • Entitlements altered
Practice	Specific management and operating rules at dam site	• Dam operation for power mazimisation • New (damaging?) communal NRM • New catchment management systems • Lack of cascade management • Poor regional HP coordination
Decision making	Catchment users, resources, skills, assets, time horizon and technology	• New resource interests • New political relationships • High development intensities • New institutional mixes • Entitlements • Shifting stategic decision making
Society	The nature of society	• Access rights to resources • Gender divisions • Corporate rights, *de facto* and *de jure* • Administrative rights, *de facto* and *de jure* • Corruption
State	The nature of the state	• Official land tenure laws • State ability to enforce law • State administrative ability • State HE, water and human rights policies • Judicial independence
Region	Regional tensions/cooperation	• Energy markets • Water allocation agreements • Regional agreements • Rate of development • Strategic interests • Donors
World	The international economy	• Market behaviour • International treaties • Interests of other countries • Oil prices

Figure 2.1 Chain of explanation applied to a hydropower site

Source: adapted from Adams, 2001.

For example, as the structure is developed at a local scale in activities such as forest clearing and resettlement, the linkages to livelihoods are transparent. Furthermore, starting at the end of the chain we can trace back how oil prices and market behaviour eventually impact local economies and ecosystems. These chains of explanation are useful in regions like the Mekong Basin, where livelihoods and the environment are deeply interlinked; where the scaling is profound; and the power asymmetries between actors considerable. When applied to hydropower development, the chain of explanation helps to demonstrate how local-level actors are influenced by macro-level processes.

While the chain of explanation has been criticized as being too hierarchical (Rangan and Kull, 2009), it does offer a starting point for discussions about power and the relationships between the scales at which it is exercised.

Blaikie and Brookfield (1987, 17) provide one of the most cited definitions of political ecology as combining 'the concerns of ecology and a broadly defined political economy'. They reasoned that 'together this encompasses the constantly shifting dialectic between society and land-based resources, and also within classes and groups within society itself' (Blaikie and Brookfield, 1987). This definition explains political ecology as an interaction between society, ecology and power, from which the art of politics is derived.

Since its growth in popularity in the late 1980s political ecology has taken on many different strands and definitions that often reflect the backgrounds and leanings of the people that employ it (cf. Zimmerer and Bassett, 2003). It has moved between structural and post-structural explanations and incorporated varying degrees of analysis on political institutions, environmental change, environmental narratives and political economy (Robbins, 2004).

Political ecology has drawn from cultural ecology, radical development geography and hazards/natural disasters research (Bryant, 1998). Political ecology also has roots in Marxist theory such as relations of production theory and peasant studies (Bryant, 1998). Radical geographers, particularly Smith (1984) and Harvey (1996), influenced political ecology's critiques of the production of nature and capitalism's role in separating nature and society. O'Connor (1988, 82) makes the useful argument that the study of political and ecological concerns arises from the 'second contradiction of capitalism': the first contradiction of capitalism is its mismanagement of labour and the second is its failure to consider a functioning environment as a condition for its perpetuation. Some prominent political ecologists have suggested that part of political ecology's rise in popularity was because it offered an opportunity to rebrand unpopular Marxist theories in the post-cold war era (Watts, pers. comm., 2010).

Neo-Marxist political ecology perspectives have used the approach to engage in debates about materialism and nature in capitalist society. Watts (2000, 257) defines political ecology as a tool 'to understand the complex relations between nature and society through careful analysis of what one might call the forms of access and control over resources and their implications for environmental health and sustainable livelihoods'. Lipietz (2000, 4) takes this definition further by stating that '[p]olitical ecology, like the Marxist-inspired workers' movement, is based on a critique – and thus an analysis, a theorized understanding – of the "the existing order of things"'. More specifically, Marx focused on the human–nature relationship, and, even more precisely, relations among people that pertain to nature (or what Marxists call the 'productive forces') (Lipietz, 2000, 70). In Marxist interpretations of political ecology, capitalism is the primary cause of environmental degradation. We follow the Marxist view of political ecology by arguing that, in the Mekong, it is capitalism's growth in China, Thailand and Vietnam (the wealthier nations of the basin) and associated burgeoning electricity demands that is one of the elements that causes environmental degradation through hydropower development that ignores its social and environmental impacts.

In the 1990s, political ecology began to take on post-structuralist approaches (Escobar, 1996; Baghel and Nusser, 2010). Post-structuralism is concerned with explanations of discourse as they change material relations. These new approaches examined unequal power relations and how they manifest themselves in environmental change across different scales (Escobar, 1996; Baghel and Nusser, 2010). Escobar (1996) argues that nature is socially constructed and carries multiple meanings (see also Harvey, 1996). Escobar (1996) posits that nature has two forms of capital: an extractive form and a postmodern form. An example of the extractive form is the selling of timber from forests, while the postmodern form centres on the commodification of nature, such as using rainforests for pharmaceuticals. 'Nature is reinvented as environment so that capital, not nature and culture may be sustained' (Escobar, 1996, 49). Using Escobar's approach, hydropower development in the Mekong Basin appears to encompass both forms of capital. Dams can be destructive in their environmental impacts on water resources, but the generation of electricity from water can also offer postmodern forms of capital.

An increasing focus on narratives and discourses further accompanied the new approaches in political ecology. Robbins (2004) argues that political ecology is both a hatchet and seed. As a hatchet, political ecology deconstructs myths, narratives and discourses linked to the control of natural resources. In the Mekong, this hatchet approach is useful for analysing the development narratives from powerful actors that state hydropower brings benefits to the whole region and those from NGOs that emphasize the 'pristine' environment. The seed is political ecology's attempt to develop new knowledge that will influence policy and natural resource management to be more equitable.

Watts and Peet (2004) link political and social movements that resist environmental change to their relationships with the state and international political economies. These 'liberation ecologies' highlight the defence of, and the struggle over, land and rights against powerful actors from government or the private sector. They argue that 'political ecology opened the possibility of a serious discussion of how nature and environmental problems were represented and how discursive formations shaped policy and practice' (Watts and Peet, 2004, 10). Political ecology began to examine how actors use social constructs of environmental problems to legitimize their positions. As Bryant (1998, 87) states, 'conflicts are … as much struggles over meaning as they are battles over material practices'. This strand of inquiry is especially important in the Mekong Basin, where terms like 'sustainable hydropower' are mercurial – there is no local consensus as to what they mean, and they can therefore mean everything or nothing (cf. Swyngedouw, 2007).

In terms of hydropower development in the Mekong, actors socially construct the impacts and benefits of hydropower across different scales and use these constructions as justification for their actions. The struggles of social and political grassroots movements against environmental change – or in the case of Laos, their absence – help to highlight who stands to win and lose from environmental change and why. Using a political ecological analysis at the meso-scale helps to understand the political struggles over participation in meaning-making and decision-making processes.

Forsyth (2003) uses a critical political ecology approach that understands the social and political influence on science as the starting point in understanding environmental change. Although all political ecology should be considered critical, Forsyth stresses that environmental knowledge and facts are constructed as part of political and economic debate. Critical political ecology considerations are important in the Mekong Basin hydropower debate. Käkönen and Hirsch (2009) point out that politics often define what and how the Mekong River Commission researches and evaluates water governance in the basin. Impact assessments that downplay the potential impacts of dams as 'benign', 'low impact' and, even, 'beneficial', and NGO documents that portray these same impacts as 'enormous', 'irreversible' and 'disastrous', also speak to a political ecology understanding of narratives and the influence of politics on science. It also speaks to how narratives form, and compete with each other, around adjectives.

Until 2000, much of political ecology's focus was in the developing world. Bryant (1998, 89), for example, argues that political ecology 'seeks primarily to understand the political dynamics surrounding material and discursive struggles over the environment in the Third World'. While Bryant and Bailey's (1997) work is useful in highlighting the role of the state and its relationship with development agencies, the value of creating a distinction between First and Third World political ecologies is put into question (Walker, 2003). Furthermore, Bryant and Bailey's (1997) text emphasizes the role of politics in shaping ecology in the Third World, but offers little in the way of specific discussion of what defines ecology. Walker demonstrates that, although some political ecologists engage with politics more than ecology, they are still concerned with ecology as an important aspect of their research. These concerns of ecology 'become primarily questions of power, struggle and representation' (2005, 78). As outlined above, Watts and Peet (2004) argue that political ecology defines and understands the term ecology through the lens of politics and as a result it introduces more perspectives than those deployed by natural scientists. Political ecology's treatment of the term 'ecology' has been adopted by other groups seeking to incorporate ecological concerns into their arguments.

Although political ecology has primarily focused on the global south, its application in the developed world context has continued to expand since the late 1990s. Atkinson attempted to link political ecology with social movements in the United States by arguing that '[p]olitical ecology is both a set of theoretical propositions and ideas on the one hand and on the other a social movement referred to as the "ecology movement" or, latterly, the Green Movement' (1991, 18). Robbins and Sharp (2003) take the application of political ecology analysis to the heart of America by analysing the moral economy of the American lawn.

Bakker (2000) used political ecology to analyse drought in England and, later in 2010 and 2004, used the approach to analyse the privatization of water in the United States, England and Wales. Work by Ubokudom and Khubchandani (2010) uses a political ecology approach to analyse health care in the United States, while Horowitz (2012) employed the approach to look at grassroots movements protesting against industry in the global north.

To illustrate political ecology's usefulness in global issues, Peet *et al.* (2011) cast a political ecological lens across issues such as over-fishing, climate change and waste. The authors develop a post-structural view of political ecology to examine how knowledge is produced and legitimized in environmental governance. By examining climate change through a political ecological lens they draw out its links to capitalism, expert knowledge, as well as to common discourses and narratives.

Political ecology on narratives

In the Mekong, narratives are often employed by politicians, industry or NGOs to frame problems and orient actions. As Roe (1991) argues, however, they can also be employed to justify interventions and marginalize or blame actors for environmental degradation. For example, neo-Malthusian arguments have often been used by policy-makers to blame indigenous people for land degradation (cf. McCann, 1999). Cronon (1995) argues that narratives are particularly powerful in environmental history because they ascribe order and agency to human-induced environmental change. Actors can frame solutions within persuasive narratives and promote narratives that are beneficial to their agendas (Leach and Mearns, 1996). Hardin's (1968) tragedy of the commons narrative has often been used by policy-makers in southeast Asia, and around the world, to remove local people from forested areas and place these under state control. Extensive evidence, however, demonstrates that common, private and state resource management options are all viable options and often politics and economics are the real drivers of resource degradation (Feeny *et al.*, 1990). Roe (1991) argues that, like Hardin's narrative, other popular narratives persist because they buttress and endorse decision-making, thereby lending authority to actions and policies.

Narratives surrounding electricity scarcity, an abundance of hydropower potential and poverty alleviation are all prevalent in the Mekong Basin. These narratives are reinforced by constructed and contested knowledge around the benefits and costs of dams. In the Mekong Basin hydropower debate, narratives have been employed by various actors to legitimize their activities and to obscure irregularities and possible negative impacts. For example, states developing hydropower often talk of benefits such as flood control and income. Conversely, international non-governmental organizations (INGOs) in the region often talk of loss of food security and broken livelihoods. Political ecology's focus on narratives has critically explored the multiple meanings of the environment and development (Watts and Peet 2004).

By understanding how power influences narratives at the meso scale in the Mekong hydropower development we can begin to understand the drivers and enablers of this development. Narratives surrounding hydropower in the Mekong Basin often emerge during the impact assessment process, as this is a key point of engagement in hydropower development. Narratives are also often framed at particular geographic scales. In the Mekong, these geographic scales can emerge as common scalar referents such as local, provincial and basin, or more political and contested ones such as the Greater Mekong Subregion or the Mekong Delta. Political ecology's analysis of scale is important to understanding narratives in that the scale at which narratives are employed is often constructed and intertwined with power relations as much as the narratives themselves.

Political ecology on scale

Understanding environment and natural resource exploitation and management between different geographical scales and how scalar constructs interlink between structures and systems is crucial to analysing hydropower development processes in the Mekong Basin. For our purposes, scale can be understood in three different contexts. First, as a physical measure of space; secondly as a space in which knowledge and power exist and change; and finally as a social construct mobilized by actors to lend credence to their agendas (Molle, 2007).

Geographers have extensively analysed the politics of scale, which can be understood as the processes within, and the emergence of, scalar constructs (Castells, 1996; Smith, 1984; Swyngedouw, 2001); or, indeed, 'disconnects' (Suhardiman *et al.*, 2012). Gandy (2002) in his examination of New York City analysed how political economic processes shaped the 'nature' of the city. Political ecology examines the production and operation of the politics of natural resource allocation at regional and national scales. Gandy's work discusses how the scale of the city is constructed and how nature has been constructed within this scale. Escobar (1996) further argues that capital redefined nature as an 'environment'. This new definition ultimately dismisses ideas of nature and culture, so that the power of those who control capital can maintain their status. Lefebvre (1990) highlights how geographical spaces and their scalar concepts should be considered as socially produced. Much of the geographic literature on scale and space link them to the survival of capitalism and its expansionist tendencies. Geographic literature provides a foundation on which political ecology theory and its analytical approach has advanced the understanding of scale.

Zimmerer and Bassett (2003) argue that political ecology is concerned with how geographic scale is constructed and how actors use it to legitimize or delegitimize environmental change. Political ecology analysis argues that environmental and social change is the result of how the processes and mechanisms between local, national and international levels interact (Marston and Smith, 2001; Rangan and Kull, 2009). Actors and actions at one scale may have impact on activities at another. As Bryant and Bailey (1997, 33) argue, 'different actors contribute to, are affected by, or seek to resolve, environmental problems at different scales'. Political ecology can help to illuminate how actors at one scale can construct and disseminate a version of scale that advances their own agendas. Political ecology can help to highlight how powerful actors define the scale of environmental change and use these definitions to legitimize their actions. In the Mekong Basin hydropower debate, hydropower developers and states often claim that the scalar benefits of hydropower, which are often measured at meso or national scales, are larger than the costs, which are often measured at local scales. For example, Sneddon and Fox (2006) contend that the World Bank constructed the benefits of the Pak Mun Dam around an economic development narrative (i.e. that the dam would benefit the country as a whole) and de-emphasized the local costs of the dam on fisheries and livelihoods.

Rangan and Kull (2009) highlight the different ways scale is constructed to explain ecological and social change. A political ecology analysis shows that scale is socially constructed and that the politics of scale are defined through institutions, events, technologies, politics and measurements. In the Mekong Basin, the benefits and costs of hydropower development are dependent on definitions of scale. The politics surrounding the definition of scales are fundamental to identifying the winners and losers in the current phase of hydropower intensification. A political ecology analysis of how scales interact can be used to locate environmental change in political economic systems and state relations (Paulson and Gezon, 2005). For example, in Mekong hydropower development, environmental change is often debated at a meso scale between states and unequally balanced against economic growth.

Political ecology has a long tradition of detailed local-scale case studies that link to national-scale politics and economics that links local-level surveys to global agents. For example, political ecology uses the chain of explanation to examine the structures between scales, often concluding that national or international politics and economics influence local environmental change.

Towards a political ecology of Mekong hydropower development

Early uses of political ecology (i.e. before the late 1990s) did not address water resources issues. In 1997, Bryant and Bailey argued that political ecologists have been 'relatively negligent' in analysing water, hydropolitics and 'how control over water is linked to unequal power relations' (1997, 193). It was not until the late 1990s that political ecology began to analyse water issues with useful results (Swyngedouw, 1997; Bakker, 1999; Sneddon and Fox, 2006). Because water is so politicized, political ecology can be useful in analysing issues related to the use, distribution and conflict over water.

Political economy has also been used to examine water governance and hydropower development. Mitchell (1998) and Radosevich and Olson (1999) use political economy arguments to analyse the past half century of water use and hydropower development. These researchers remind us that power and economics in water management are not as straightforward as the large state versus the homogeneous community. Mitchell's (1998) analysis of the political economy of Mekong Basin development focuses on the political aspects of decision-making and how the benefits and impacts of decisions have uneven distributions. Mitchell also begins to analyse some of the geographical scales and their linkages, looking at global influences on development and how these translate into changes in regional and local scales. Importantly, Mitchell identified the emergence of conflict between basin-wide coordination, promoted by the Mekong River Commission (MRC) and its donors, and the individual national agendas. The conflicting agendas of the MRC and individual basin states gained pace during the 1990s as the region stabilized and financing availability increased. Mitchell's (1998, 79) statement that the MRC is 'constrained by the political environment within which it functions' is as true today as it was 20 years ago.

As Bakker argues, political ecology offers a more nuanced perspective than political economy on water issues because the approach 'acknowledges the materiality of nature' thereby, 're-theorizing resource regulation; and interrogating the role of the state from a different perspective than that of much political economy' (Bakker, 2010, 52). In other words, political ecology critically examines the process of resource commodification and regulation. Bakker further states that, by 'acknowledging the coproduction of socio-economic and environmental change', political ecology helps to 'generate new insights into contested and complex periods of transition' in water governance.

The utility or value of the approach to understanding the political and economic landscape and then identifying leverage points for change becomes apparent when political ecology is applied to Mekong Basin water governance and hydropower development. First, in the Mekong Basin, a complex period of transition between modes of regulation has emerged. The transition has involved a shift from *state and development bank* to *private sector hydropower development*. Bakker states that 'The quotidian practices of regulation develop within and reinforce but also sometimes contradict broader macro-economic patterns of resource regulation' (2010, 53). As the actors and regulators transition to the private sector mode in the Mekong Basin, new political and economic mechanisms have emerged while others have gained influence or disappeared. The political ecology approach is useful in analysing the political and economic forces that drive and enable these transitions. For example, the World Bank and the Asian Development Bank, in line with neo-liberal trends, encouraged private-sector investment in the basin. This private-sector investment brought with it new requirements for confidentiality which has resulted in less transparency and greater obscurity.

Swyngedouw (1999) uses a political ecology approach to demonstrate how nature and society are meshed together. Many people living in the Mekong Basin depend daily on the fish they catch and non-timber forest products (NTFPs). Economies and livelihoods in the basin are geared to the flood pulse of the river. For example, 80 per cent of the animal protein in the Cambodian diet is from wild fish resources (Van Acker, 2003). In many places where opposition to dams has occurred it has arisen because of concerns that local fisheries might collapse or that water quality and quantity might decline. For many people in the Mekong, nature and society could be considered as inseparable. In order to understand and analyse water governance in the Mekong it is necessary to examine both political and ecological factors.

Using a political ecology approach to understand environmental change in Bangladesh, Bradnock and Saunders (2000, 67), usefully point out that 'politics itself is not self-explanatory or uncontested'. Environmental change occurs within interlinked combinations of environmental and human systems (Bradnock and Saunders, 2000). In Laos, arguably more than many developing countries, environmental and human systems are deeply intertwined and each influence environmental change in significant ways. This volume uses a political ecology approach that focuses on political, social and economic forces to understand the drivers and enablers of hydropower development because hydropower is developed to serve political and economic systems.

By including the principles of both social and environmental justice, political ecology can offer a more in-depth understanding of the winners and losers in the current hydropower intensification in the Mekong. As Bakker writes, this is because political ecology 'begins from the assumption that socio-economic and environmental change are co-produced, but also broadens the set of actors – non-humans, as well as humans – who are considered both as objects of study, and also as holders of legitimate claims to equitable treatment' (2010, 54). An effective analysis of the processes of evaluation and implementation of Mekong hydropower projects requires consideration of hydropower's impacts and advantages on the regional and international actors and the environment.

Finally, political ecology may provide an additional and complementary analytical lens when it comes to international relations. In the Mekong Basin, the private power producer model has changed the role and relationships between various states, the nature of the market and citizen relations, and the allocation and use of water and other natural resources. By acknowledging the role of the state in resource management, a political ecology approach helps to understand both state and private-sector motivations, shifts in responsibilities and the nature of the evolving relationships (Bakker, 2010).

A Sneddon and Fox (2006) paper on the critical hydropolitics of the Mekong is a strong example of the application of the political ecology at the basin scale. They demonstrate how in many areas of the world, and particularly in the Mekong, water is a resource that is managed by elites. They argue that political ecology helps to identify how actors represent rivers and how institutions and systems legitimize these representations. They propose that such contextual analysis has helped to illuminate the 'dark corners' in previous transboundary hydropolitical studies. They remind us that cooperation in river basins can lead to exploitation of natural resources and the people who depend on them.

Conclusion

Political ecology is, then, a powerful tool with which hydropower, its development, the ways its costs and benefits are shared, the resettlement it prompts and the politics that surround it can be analysed. Others before us have perceived this utility, and have sought to unpack the often-unseen world that lies behind hydropower decision-making. Most political ecological research in the region has tended towards asymmetries of power and the politics of water. The MRC has received more attention from political ecologists than perhaps it deserves, given its relatively minor role in the region's hydropower decision-making processes. A modest body of work has focused on narrative construction (Bakker, 1999; Wong, 2010), and 'water grabbing' (Matthews, 2012).

There remains considerable scope for additional research political ecological on the Mekong's hydropower development and its water resource development, providing a robust theoretical basis from which to analyse the region's hydropolitical dynamics. In this chapter, we have sought to demonstrate how this theoretical basis can help to uncover and explain these processes within the region. The approach has great validity considering the environmental and political changes and trends of the region; and sufficient flexibility to accommodate the local, national, regional and global trends and processes that shape hydropower decision-making here.

The chapters contained in this volume all seek to demonstrate the approach's applicability to understanding the region's hydropower dialectics and politics, and, indeed, its relationship with nature.

References

Adams, W. (2001) *Green Development: Environment and Sustainability in the Third World*, 2nd edn, London: Routledge.

Atkinson, A. (1991) *Principles of Political Ecology*, London: Belhaven.

Baghel, R., and Nusser, M. (2010) 'Discussing large dams in Asia after the World Commission on Dams: Is a political ecology approach the way forward?', *Water Alternatives*, 3(2): 231–48.

Bakker, K. (1999) 'The politics of hydropower: Developing the Mekong', *Political Geography*, 18: 209–32.

Bakker, K. (2000) 'Privatizing water, producing scarcity: The Yorkshire drought of 1995', *Economic Geography*, 76(1): 4–27.

Bakker, K. (2004) *An Uncooperative Commodity: Privatizing Water in England and Wales*, Oxford: Oxford University Press.

Bakker, K. (2010) *Privatizing Water: Governance Failure and the World's Urban Water Crisis*, Ithaca, NY: Cornell University Press.

Blaikie, P., and Brookfield, H. (1987) *Land Degradation and Society*, London: Methuen.

Bradnock, R. W., and Saunders, P. (2000) 'Sea-level rise, subsidence and submergence: The political ecology of environmental change in the Bengal delta', in P. Stott and S. Sullivan (eds), *Political Ecology: Science, Myth and Power*, London: Edward Arnold, pp. 66–90.

Bryant, R. L. (1998) 'Power, knowledge and political ecology in the third world: A review', *Progress in Physical Geography*, 22(1): 79–94.

Bryant, R. L., and Bailey, S. (1997) *Third World Political Ecology*, London: Routledge.

Castells, M. (1996) *The Rise of the Network Society*, Malden, MA: Blackwell Publishers.

Cockburn, A., and Ridgeway, J. (eds) (1979) *Political Ecology*, New York: Times Books.

Cronon, W. (1995) *Uncommon Ground: Toward Reinventing Nature*, New York: Norton.

Dahl, R. (1957) 'The concept of power', *Behavioral Science*, 2(3): 201.

Ehrlich, P. R. (1968) *The Population Bomb*, New York: Buccaneer Books.

Escobar, A. (1996) 'Construction nature: Elements of a post-structural political ecology', *Futures*, 28(4): 325–43.

Feeny, D., Berkes, F., McCay, B., and Acheson, J. M. (1990) 'The tragedy of the commons: Twenty-two years later', *Human Ecology*, 18(1): 1–19.

Forsyth, T. (2003) *Critical Political Ecology: The Politics of Environmental Science*, London: Routledge.

French, J. R. P., and Raven, B. (1959) 'The bases of social power', in D. Cartwright and A. Zander (eds), *Group Dynamics*, New York: Harper & Row, pp. 150–67.

Gandy, M. (2002) *Concrete and Clay: Reworking Nature in New York City*, Cambridge, MA: MIT Press.

Hardin, G. (1968) 'The tragedy of the commons', *Science*, 162: 1243–8.

Harvey, D. (1996) *Justice, Nature and the Geography of Difference*, Malden, MA: Blackwell Publishers.

Horowitz, L. S. (2012) 'Power, profit, protest: Grassroots resistance to industry in the global north', *Capitalism Nature Socialism*, 23(3): 20–34.

Käkönen, M., and Hirsch, P. (2009) 'The anti-politics of Mekong knowledge production', in F. Molle, T. Foran and M. Käkönen (eds), *Contested Waterscapes in the Mekong Region: Hydropower, Livelihoods and Governance*, London: Earthscan, pp. 333–55.

Leach, M., and Mearns, R. (eds) (1996) *The Lie of the Land: Challenging Received Wisdom on the African Environment*, Oxford: James Currey.

Lefebvre, H. (1990) *The Production of Space*, London: Basil Blackwell.

Lipietz, A. (2000) 'Political ecology and the future of Marxism', *Capitalism, Nature, Socialism*, 11: 69–85.

McCann, J. C. (1999) *Green Land, Brown Land, Black Land: An Environmental History of Africa, 1800–1990*, Oxford: James Currey.

Marston, S. A., and Smith, N. (2001) 'States, scales and households: Limits to scale thinking? A response to Brenner', *Progress in Human Geography*, 25(4): 615–19.

Matthews, N. (2012) 'Water grabbing on the Mekong Basin: An analysis of the winners and losers of Thailand's hydropower development in Lao PDR', *Water Alternatives*, 5(2): 392–411.

Miller, A. (1978) *A Planet to Choose: Value Studies in Political Ecology*, New York: Pilgrim Press.

Mitchell, M. (1998) 'The political economy of Mekong Basin development', in P. Hirsch and C. Warren (eds), *The Politics of Environment in Southeast Asia: Resources and Resistance*, London: Routledge, pp. 71–89.

Molle, F. (2007) 'Scales and power in river basin management: The Chao Phraya River in Thailand', *Geographical Journal*, 173(4): 358–73.

O'Connor, J. (1988) 'Capitalism, nature, socialism: A theoretical introduction', *Capitalism, Nature, Socialism*, 1: 11–38.

Paulson, S., and Gezon, L. (eds) (2005) *Political Ecology across Spaces, Scales, and Social Groups*, Brunswick, NJ: Rutgers University Press.

Peet, R., Robbins, P., and Watts, M. J. (eds) (2011) *Global Political Ecology*, London: Routledge.

Radosevich, G., and Olson, D. (1999) 'Existing and emerging basin arrangements in Asia: Mekong River Commission case study', 3rd Workshop on River Basin Institution Development, Washington, DC: World Bank, 24 June.

Rangan, H., and Kull, C. A. (2009) 'What makes ecology "political"? Rethinking "scale" in political ecology', *Progress in Human Geography*, 33(1): 28–45.

Robbins, P. (2004) *Political Ecology: A Critical Introduction,* Oxford: Blackwell Publishing.

Robbins, P., and Sharp, J. T. (2003) 'Producing and consuming chemicals: The moral economy of the American lawn', *Economic Geography*, 79(4): 425–51.

Roe, E. (1991) 'Development narratives or making the best of blueprint development', *World Development*, 19(4): 287–300.

Russett, B. (1967) *International Regions and the International System: A Study of Political Ecology,* Chicago, IL: Rand McNally.

Smith, N. (1984) *Uneven Development: Nature, Capital and the Production of Space*, Oxford: Blackwell Publishers.

Sneddon, C., and Fox, C. (2006) 'Rethinking transboundary waters: A critical hydropolitics of the Mekong Basin', *Political Geography*, 25: 181–202.

Suhardiman, D., Giordano, M., and Molle, F. (2012) 'Scalar disconnect: The logic of transboundary water governance in the Mekong', *Society and Natural Resources*, 25: 572–86.

Swyngedouw, E. (1997) 'Power, nature, and the city: The conquest of water and the political ecology of urbanization in Guayaquil, Ecuador, 1880–1990', *Environment and Planning A*, 29(2): 311–32.

Swyngedouw, E. (1999) 'Modernity and hybridity: Nature, regeneracionismo, and the production of the Spanish waterscape, 1890–1930', *Annals of the Association of American Geographers*, 89(3): 443–65.

Swyngedouw, E. (2001) 'Neither global nor local: "Glocalization" and the politics of scale', in B. Jessop (ed.), *Regulation Theory and the Crisis of Capitalism*, Cheltenham and Northampton, MA: Edward Elgar, pp. 137–66

Swyngedouw, E. (2007) 'Impossible/undesirable sustainability and the post-political condition', in J. R. Krueger and D. Gibbs (eds), *The Sustainable Development Paradox*, New York: Guilford Press, pp. 13–40.

Ubokudom, S. E., and Khubchandani, J. (2010) 'The ecology of health policymaking and reform in the USA', *World Medical and Health Policy*, 2: 1–32.

Van Acker, F. (2003) *Cambodia's Commons: Changing Governance, Shifting Entitlements?,* Discussion Paper 42, Antwerp: Centre for ASEAN Studies, University of Antwerp.

Walker, P. A. (2003) 'Reconsidering "regional" political ecologies: Toward a political ecology of the rural American West', *Progress in Human Geography*, 27(1): 7–24.

Walker, P. (2005) 'Political ecology: Where's the ecology?', *Progress in Human Geography*, 29(1): 73–82.

Watts, M. (2000) 'Political ecology', in T. Barnes and E. Sheppard (eds), *A Companion to Economic Geography*, Oxford: Blackwell, pp. 257–75.

Watts, M., and Peet, R. (2004) 'Liberating political ecology', in R. Peet and M. Watts (eds), *Liberation Ecologies: Environment, Development, Social Movements*, 2nd edn, London: Routledge, pp. 3–47.

Wolf, E. (1972) 'Ownership and political ecology', *Anthropological Quarterly*, 45: 201–5.

Wong, S. M. T. (2010) 'Making the Mekong: Nature, region, postcoloniality', DPhil Dissertation, Ohio State University, Columbus.

Zimmerer, K. S., and Bassett, T. J. (2003), 'Approaching political ecology: Society, nature, and scale in human-environment studies', in K. S. Zimmerer and T. J. Bassett (eds), *Political Ecology: An Integrative Approach to Geography and Environment-Development Studies*, New York: Guilford Press, pp. 1–25.

3 A political ecology of hydropower development in China

Zha Daojiong

Introduction

China is home to the world's largest number of dams, largest amount of hydropower generation, largest irrigated area, largest hydropower project and largest water transfer project. It remains increasingly active in hydropower investment both at home and abroad, in spite of the growing number of controversies and criticisms. This chapter will offer a political ecology perspective on these trends and preview the drivers, both ideational and institutional, that sustain the country's hydropower policy choices. It will shed light on how the Chinese react to international commentaries regarding their performance.

The chapter begins with a brief description of political ecology as an analytical tool to examine Chinese hydropower policy. The second section reviews the interests and actors that sustain hydropower development within the country. In the third section, the chapter considers Chinese justifications for involvement in hydropower development in the Lower Mekong region (LMR). Our concluding observations advocate driving home the necessity for a multiple stakeholder approach to Chinese investments in the LMR by reaching out to Chinese corporate executives and jointly developing project-specific standards for improved protection of environmental and human welfare.

Applying 'political ecology' to understanding Chinese hydropower development

Electricity is a non-storable but critical input for economic growth. Studies have found that 'a 1% increase in electricity generation growth would raise China's GDP growth by about 0.6%' (Cheng *et al.*, 2013, 369). Despite the rapid expansion of China's electricity infrastructure, electricity shortages remain common. For example, in 2011, 11 out of the country's 31 provinces suffered from an electric power shortage, ranging from 5 per cent to 19 per cent (Zeng *et al.*, 2013, 611).

In 2012, hydropower accounted for about 16 per cent of China's electricity production and 7 per cent of its total energy consumption. With coal-fired power still accounting for 80 per cent of China's power generation, Chinese government policies encourage further exploitation of hydropower. In part, this is motivated by the necessity to address power shortages, and in part it is because hydroelectricity as a replacement for coal-fired power generation brings with it positive impacts on the abatement of air pollution.

As is well known, China's hydropower projects, particularly the large ones, have caused many problems, including the degradation of freshwater, soil and river erosion, and increasingly expensive and contentious population resettlements (Mertha, 2008). On the one hand, protests by affected populations and non-governmental associations against hydro-dam projects have motivated the central government to shelve industry plans for developing river cascades such as on the Nu River (Magee and McDonald, 2006). On the other hand, further growth in hydropower remains a viable option in virtually all of China's energy policy plans.

Political ecology is an approach to understanding complex nature/society relationships. It critically examines popular narratives about the pursuit of economic development by raising issues of justice, society and nature in specific contexts (Bryant and Bailey, 1997; Geheb and Mapedza, 2008; Matthews and Schmidt, 2014). Scholars have found the approach particularly useful for appraising the pursuit of hydropower as a solution for low-income societies, including those in the LMR (e.g. Yong and Grundy-Warr, 2012).

Applied to the study of contemporary China, political ecology has been used to make sense of the management of the environmental consequences that derive from but are not exclusively caused by the pursuit of hydropower. Some scholars place more emphasis on the role of ideology, arguing that the long shadow of the Maoist pursuit of rapid industrialization for the sake of industrial independence is a cause of these inadequacies (Economy, 2005). Other scholars take note of the progress made in nature conservation but stress the importance of being more attentive to civil society voices that advocate adequate protection of ecological and individual interests (Xu and Melick, 2007). In this line of inquiry, it is not so much that China disregards the utility of laws and rules, but rather that effective enforcement is what truly matters (Tilt, 2007). Others highlight the continuation of the top–down paradigm of public policymaking in China that helps to retain overwhelming administrative power in the hands of central government bureaucracy. By extension, market forces can and should be allowed to play a bigger role in affecting positive change on the ground (Zeng *et al.*, 2013).

Each of these angles is relevant in appraising China's water governance in general and the pursuit of hydropower in particular. In China, hydropower is promoted by the government as more than just a form of energy to satisfy present-day demand. It is a reference point of national pride – in the sense of demonstrating resolve when faced with adversarial conditions. An illustrative case is the country's response to the collapse of Banqiao, Shimantan and a series of smaller dams (in Henan Province) in August 1975, perhaps the worst ever recorded disaster in world history if measured by the loss of human life (Sovacool, 2008). China treated this as an opportunity to upgrade its indigenous dam construction engineering capacity. Its sense of 'pride' is such that China managed to improve its dam construction and operation safety record. Indeed, there has been no major dam failure since then. Of course, this does not imply that dam safety in China can be assured in the future.

The official Chinese narrative for justifying the further pursuit of hydropower now includes the reduction in greenhouse gas (GHG) emissions. China finds it convenient to claim that increasing its hydropower capacity shows that the country is a responsible nation, notwithstanding its refusal to agree on an internationally binding emissions reduction target (Conrad, 2012). It needs to be noted, meanwhile, that the claim that hydropower is a clean source of energy is subject to scientific contestation. Some claim that it disregards the massive amounts of fossil fuels consumed in producing the industrial materials used for dam construction, together with the reservoirs, which contribute to further GHG emissions. Yet others argue that the extent to which reservoirs contribute to GHG emissions is not clear-cut. Some reservoirs continue to give off large amounts of GHG, while others do not (Tortajada *et al.*, 2012). Merits and demerits in the dam–GHG debate are beyond the scope of this chapter. Our point here is that, at the level of government policy, China has chosen to cite a global concern as a means to pacify the voices of opposition, especially against the background of coal-fired electricity that still dominates the country's power supply.

A third dimension of the prevailing official Chinese narrative is that China's involvement in dam construction and hydropower development in developing countries, which is often through government-funded development aid, is conducive to poverty reduction. Indeed, against the background of traditional funders such as the World Bank choosing to reduce dam project funding in the LMR in the late 1990s, China stepped in to fill the void (Middleton *et al.*, 2010), and it continues to do so, both through government-funded projects and privately funded ones.

In any case, the reality is that the demand for energy in China is set to grow further, to feed its rapidly urbanizing population and overall economic growth. Standard Chinese projections see 2030 as the year for China's energy demand to peak (Shan *et al.*, 2012). Also set to increase is the Chinese interest in, and capacity for, exporting industrial equipment and products, together with its engineering expertise in hydropower dam construction and operation.

To be sure, economic and energy policy-makers in China as well as their critics are well aware of the profound challenges of managing the country's water stress: a deficient northern region and surplus southern region, plagued with all sorts of water-related disasters, from drought to flooding. As is well documented, the official Chinese discourse demonstrates a determination to pursue 'sustainable' strategies of development that are consistent with the preservation of the environment. In contrast, the global discourse on hydro engineering is much more nuanced. Man-made projects to manage water flows in large river systems such as the Rhine and the Mississippi, the Nile, the Murray-Darling and the Oxus went on for much of the twentieth century. But questions remain over whether they are sustainable after all (e.g. McCormack, 2001).

Major dam/hydropower projects in China, most notably the Three Gorges Project (TGP), have attracted extensive research attention among Chinese scholars. Indeed, it could be said that the TGP has become an icon in the debate over the wisdom of hydropower in China. Before the TGP was implemented, Chinese literature opposing dam development tended to employ mega-narratives arguing against its necessity (cf. Dai, 1998). Now that the TGP has become a reality, Chinese scholars choose to focus on more tangible issues, such as dealing with the environmental consequences (cf. Tan and Yao, 2006) as well as giving more attention to the resettled population (cf. Xi and Hwang, 2011).

Over the years, there has been no shortage of effort made by international development agencies to assist China to manage its hydropower projects. These efforts include project funding (Martinot, 2001), capacity building (Heggelund, 2004) and numerous collaborative research programmes with Chinese hydropower engineering and policymaking professionals. In other words, professional advice from abroad has been well presented to Chinese decision-makers. Admittedly, the degree of impact of such international expert advice is difficult to gauge.

Therefore, what makes Chinese water bureaucrats demonstrate continued strong faith in pursuing water engineering, irrigation, flood control, and dam building projects, both inside and outside their own country? It is particularly important to answer questions like this against a background of China seeking international acceptance of its approach to development. For example, we can ask, with major international development institutions such as the World Bank and the Asian Development Bank placing and refining environmental and social safeguards in their funding for hydropower projects, are Chinese government agencies, state-owned banks and state-owned hydro engineering companies learning from international practices?

In the rest of this chapter, we will analyse the Chinese political ecology of hydropower development by considering the variety of actors both in and associated with the hydropower industry. By trying to understand their positions and interests, we will remind the readers of the various opinions and forces that have sustained the pursuit of hydropower as a viable option in China, together with the Chinese involvement in hydropower development in the LMR.

Interests that sustain the pursuit of hydropower

As is true of any country or society, interests have multiple dimensions, and both ideational and institutional traits. By 'ideational' this chapter refers to those articulations that are meant to galvanize the competing views towards acquiescence over a policy position that reflects one participating party's self-interests. References to 'institutions' in this chapter include only those formal governmental agencies whose relationships can provide a cumulative understanding of hydropower decision-making.

This is not to say that Chinese non-governmental organizations (NGOs) do not feature in China's hydropower decision-making. On the contrary, Chinese

NGOs, often through allying with their international peers, have demonstrated their potency in putting a stop to some of the controversial large hydropower and water works development plans (Morton, 2007). Advocacy certainly plays a role in China, and has resulted in a number of planned dams not being built (Mertha and Lowry, 2006). But as is true of so many other facets in the pursuit of economic growth in China, those institutions that make up the formal organs of the state continue to be in the driving seat (Huang and Yan, 2009).

This chapter examines the Chinese governing agencies at the local and district, provincial and national levels. As part of this analysis, the chapter further recounts China's utilization of bilateral and multilateral actors in pursuing its hydropower development projects both within and beyond its own nation-state boundaries.

Drivers for hydropower development in China

There are multiple drivers for Chinese hydropower development. At the ideational level, for centuries the country has faced three persistent water-related challenges: flooding, water scarcity and water pollution. The apparent Chinese faith in the power of engineering is rooted in deep and persistent adherence to Confucian philosophies (Economy, 2002, 2005). As already mentioned, Chinese government energy policy presents hydropower as 'clean' energy, necessary for reducing the country's GHG emissions and its reliance on coal. Public demand for improved air quality in urban areas reinforces and legitimizes the pursuit of hydropower.

At the institutional level, all of the actors in the hydropower industry chain, from the Communist Party and its government bureaucracies to state-owned banks and firms (engineering, construction, procurement and management) are influential in sustaining the pursuit of hydropower both inside and outside the country (McDonald *et al.*, 2009). Together, these actors play the central role in implementing the country's energy development plan: non-fossil fuel to account for 15 per cent of primary energy consumption by 2020, compared with the 2005 level (Zhao *et al.*, 2012).

International actors in China

Domestic actors are, meanwhile, not the only drivers of hydropower development in China. An episode in the country's hydropower policy reform process that has failed to receive virtually any mention in international commentaries is the so-called 'Lubuge shock'.[1] The Lubuge Hydropower Station is small, with an installed capacity of just 6 MW. The dam is situated in the mountains bordering Yunnan and Guizhou provinces, in the Pearl River Basin. Although finally completed in 1991, the project was first conceived in the mid-1960s. In 1982, a year after China gained membership of the World Bank, the Chinese government successfully secured the bank's funding for the Lubuge Project. Along with the funding came open and international bidding for the construction, procurement, engineering and management process, for the first time in China's history of hydropower development. It was also the first time that China had made use of international financing for a hydropower project.[2] The 'shock' to the Chinese hydropower industry came from the fact that a foreign consortium prevailed in the bidding process. Chinese managers and engineers had to play a supportive role according to international standards and norms. This experience opened the sector up to international standards, and led to increased innovation for all actors in China's hydropower industry.

The Lubuge Project ushered in an era of foreign actors – banks, engineering companies, power equipment producers, etc. – participating in China's hydropower development (Martinot, 2001). In the wake of the controversies associated with the international financing of the TGP in the mid-1990s, financing from abroad for hydropower in China started to decline (McGregor, 1993). Within China, concerns about the debt burden also grew due to the long lifecycle of international loans (Ma *et al.*, 2006). There has yet to be comprehensive accounting of the foreign involvement in China's hydropower sector. According to Chinese industry insiders,

> international investors generally demand their Chinese partners keep their hydro-related activities in China away from media exposure out of fear of their corporate images being made negative. [... Yet,] in the past thirty years advance in Chinese hydropower engineering and management capacities would not have been as fast if not for the continuous participation of foreign hydro companies.[3]

China's hydropower governance

Legal framework

Over the years, hydropower has been incorporated into national law, albeit only to a limited degree. The key legal instrument that accommodates the general principles of how water resources are managed is the Water Law, first promulgated in 1988 and then revised in 2002 (Wouters *et al.*, 2004). In this 82-article law, only one article (no. 26) is specifically devoted to hydropower construction:

> The state encourages the development and utilization of hydro-energy resources. On rivers with high hydropower potential hydro-energy, multi-purpose and cascade development should be promoted in a planned manner ... In the construction of hydropower stations, attention should be paid to the protection of the ecological environment, and consideration shall, at the same time, be given to the needs of flood control, water supply, irrigation, navigation, bamboo and log rafting, fisheries, etc.

> (PRC, 2004)

Associated with this general law are separate laws for flood control (1997), environmental impact assessment (2002), water pollution control (2008) and water soil conservation (2010). There is yet to be a separate law for the specific purpose of regulating hydropower development. Our research team's extensive literature review led us to a three-volume compilation of *Chinese Laws and Rules for Water Works and Hydropower* (Compilation Board, 2012). In this collection, 25 administrative laws are listed as being relevant to governing the pursuit of hydropower.

In the Chinese context, the purpose of an administrative law to govern an industry or sector is to correct the excesses of state administrative powers by institutionalizing the division of the rights and responsibilities of the administrative, corporate and (though limited) individual actors (Chen, 1998; Lubman, 2012). In practice, an administrative law is less demanding than a general law. This leaves a great deal of space for local government officials to justify their decisions to support hydropower projects under their respective geographical jurisdictions by pointing to the need for effective administration, which often means pursuing growth in the energy sector and overall economic growth as the only priorities.

Through our interviews with experts in the Development Research Centre of the Ministry of Water Resources[4] and the Chinese Hydraulic Engineering Society,[5] we learned that streamlining the hydropower decision-making process continues to rank low on the agenda, in large part because incidents and accidents associated with massive dam/dyke failures have over the years become less frequent and are therefore considered inconsequential – in economic and humanitarian terms – compared with flooding, drought and pollution. A deeper structural reason is that China has

> never quite resolved the decades-long debate over which water function should be given priority: energy development or water use for agricultural and urban use. This adds to confusion over a delineation of authority in planning and executing hydropower dams.[6]

Testimony to this debate is the history of reshuffling in the water and power industries up until the late 1980s. After the reshuffling, in theory, government administration was separated from energy project construction and operation. Although separate, government ministries and state-owned hydro companies were beholden to the same laws and legislative rules. Yet, since there is no law in place for the purpose of governing hydropower project construction, government ministries and state-owned energy companies are in constant competition for influence in the approval process of a major new hydro-dam/power project. The result is that more often than not the power/interest of a state-owned company prevails.

Who decides?

China is experiencing a burgeoning civil society movement that emphasizes the need for effective environmental protection. Large hydropower dam projects have received a growing level of attention from civil society (Xu and Melick, 2007). Yet, coupled with the domineering role that government agencies are allowed to play in hydropower project decision-making, civil society often only learns about new dam construction once it is well under way. This limits the civil society's capacity to impact decision-making by enabling public debate before a decision is made.

In today's context, who decides on a new hydropower dam project? Our research found that identifying which government agencies are tasked with managing water-related projects is much easier than understanding how decisions are made and who makes them. In a formal sense, China put in place a clear delineation of authority for managing the country's river systems, including installation of river basin commissions, as far back as the 1940s. At all four levels of government, there is an established water-related bureaucracy. The country's energy bureaucracies, meanwhile, also have a say in deciding on a hydropower project.

Tracking the flow of capital would be an ideal method for mapping the interplay between the various actors in the Chinese hydropower industry. Until the late 1970s, the only mode of finance was through the government. In the early 1980s, sources of financing started to diversify. A simplified model has the hydropower company functioning as an investing party, with the government providing subsidies and banks offering loans. As of the mid-2000s, contributors to project financing came from both central and local governments, hydropower construction and operating companies – many of which are listed on stock exchanges – loans from domestic and international banks, and proceeds from sales to the electricity grid (Luo, 2007). During the 11th five-year plan period (2006–10), the bulk (30 per cent) of financing came from the construction and operating companies. Because the government sets on-grid hydro-electricity prices lower than those for coal-fired electricity, by a 15 per cent margin in some instances, the government has to offer subsidies to keep the hydropower operations going (Luo, 2007, 70). As has been documented in industry analysis, China's on-grid electricity tariff setting continues to go through constant change (Ma, 2011).

Foreign direct investment (FDI) in Chinese hydropower sector

As stated previously, beginning with the Lubuge Project, Chinese hydropower engineers and project managers access international experience in search of 'smarter' hydropower projects. FDI in China's power sector has played its role (Blackman and Wu, 1999). But the share of FDI is difficult to ascertain. International development institutions have done their part in professionalizing China's energy projects – hydropower included – especially in the 1980s and 1990s. These 'projects helped accelerate the development of large-scale efficient coal power plants, hydropower, state-of-the-art technologies for controlling power-plant emissions, and international best practice environmental assessments of energy projects' (Martinot, 2001, 581). FDI interests make it possible for the various domestic interests to justify the continuous pursuit of hydropower as being part of an international norm.

Chinese hydropower interests have found a partner in the United Nations' Clean Development Mechanism (CDM). China began to make use of the CDM shortly after its formal launch in February 2005. As of mid-2010, China's share in worldwide CDM transactions stood at 40 per cent. Over 900 of China's 2,100 CDM projects (44 per cent) are hydropower projects (Xiong *et al.*, 2010). As can be expected, Chinese energy policy researchers take pride in tallying up CO_2 reductions in China through hydropower. As noted by an official from the Energy Research Institute:

> It is ironic for foreigners to make negative commentaries about the continuation of hydropower development in China. In reality, some Western countries clearly benefit from the sale of equipment and accompanying loans to China, while counting such CO_2 reduction as being to their credit, as agreed in the 1997 Kyoto Protocol on climate change.[7]

Interests in China that stand to benefit from further hydropower development also include the equipment manufacturers. China's hydro-geology is such that the regions with remaining potential for large-scale hydropower dams are in the country's southwest, whereas demand for electricity in eastern regions of the country continues to grow, and they need to diversify away from their current reliance on coal-fired power. The distances involved require innovation in electricity transmission, which is essential for justifying hydropower projects in the high mountains of southwestern China. A case in point is the Xiangjiaba–Shanghai Demonstration Project, which uses a high voltage direct current (HVDC) transmission line to enable the transport of hydropower out of the Jinsha River system in southwestern China for use in Shanghai. Completed in 2009, the HVDC line was the first of its kind in the world, capable of carrying 6400 MW electricity, at ±800KV, over 2034 km (Zhou *et al.*, 2010). It goes without saying that such projects easily fit into the government and industry's pursuit of status and profit from manufacturing internationally.

Over the years, small hydro projects have been viewed as socio-politically more acceptable than large-scale ones. In 2010, China installed 45,000 small hydro projects to produce 160TWh of electricity, making a significant contribution to rural electrification (IRENA, 2012, 12). Still, many remote rural areas in China have little prospect of accessing grid-based electricity, which usually only extends to the densely populated urban areas, where a large customer base justifies the electricity infrastructure investment. One option for electrification in remote rural areas is to decentralize electricity systems based on renewable energy sources. The pace of rural electrification in China, in spite of decades of continuous effort, is still slow because most economically disadvantaged villages are left to their own devices to attract investment by large state-owned energy companies (Yang, 2003; Wang *et al.*, 2006). This explains why the central government launched a 'Township Electrification Programme' in 2002, in the hope of rapid growth in renewable energy sources and decentralized electricity distribution. There has been progress in small hydropower, together with solar energy provision. Yet many households continue to suffer from inaccessible and unreliable electricity supplies (Shyu, 2012).

In short, the drivers in China that sustain the continuous pursuit of hydropower are multiple and each has found its own set of justifications. Although civil society voices in general have yet to feature prominently at the project feasibility stage, this does not mean that hydropower in China is free from internal debate. We report our key findings about these debates in the next section.

Chinese debates about hydropower development

Hydropower development, particularly through dam construction, is not without its share of controversies in China. The first issue is whether or not China should follow the United States in pursuing decommissioning and/or the demolition and removal of some of its dams. This topic gained interest after a senior engineer from the China Institute of Water Resources and Hydropower Research published a partial account of dam demolition in the United States from 1999 to 2003 for the sake of improving river system management. It was noted that dam demolition was gaining momentum, in spite of the continued planning of new dams. Some Chinese energy policy researchers and journalists argue that China should learn from the United States and demolish some of its dams (Yang, 2004). A deputy chief executive officer of the Three Gorges Group Corporation offered a passionate rebuttal accompanied by colourful photographs in an attempt to disprove the rhetoric that demolition has become an irreversible trend in dam management in the United States. In his account, calls for China to declare its dam building era over amounted to not seeing the wood for the trees and selective presentation of facts at best, as well as demonizing the functions of dams altogether (Lin, 2005). It is also interesting to note that a report in *China Energy News* quoted Michael Rogers of the United States Society on Dams as saying, at a conference in China held in October 2012, that the decommissioning of 400 small dams in the United States over the past decade does not amount to a 'race to demolition', as there are over 80,000 water dams still in operation (Hu, 2012, 20). As can be expected, the resultant quiet consensus is to pursue 'smarter dams', i.e. dams that generate electricity but with facilities installed to minimize the impact on fish migration and the discharge of sediments. This provides another justification for Chinese hydrodam developers to claim that engineering is the key to addressing environmental and ecological controversies.

The second issue of debate is compensation for the population affected by dam construction, especially those who are immediately displaced. Although numerous dams have resulted in government-directed relocation schemes, it is the handling of the directly affected population in the Three Gorges Dam project that is most frequently cited. A simplified way to characterize the debate is not so much whether or not the affected people have the right to refuse relocation or should be offered a choice of new homes; rather, there is broad concern about the costs.

> Arguably, the single most costly – both financial and political – consequence of the Three Gorges Dam is resettlement of the affected population. It continues to cast a long shadow, preventing approval of new big dams in China, although the proposed scale is far from that of the Three Gorges.[8]

Chinese policy frameworks for compensation have gone through phases of change, making up 'an adaptive process that can be understood as pragmatic' (Wang et al., 2013, 8). It should be noted, meanwhile, that that pragmatism hinges on the Chinese government's sensitivity to large-scale social protest. Organized protest by the affected populations remains the single most powerful hedge against the pursuit of hydropower that inadequately addresses the human impact caused by resettlement.

The third issue involves the quality of the environmental and social impact assessments conducted in pursuing 'smarter' hydropower dams. Environmental impact assessments (EIAs) and social impact assessments (SIAs) are not new in China. Their formal implementation began in the mid-1980s (Wang et al., 2003). Chinese hydropower dam regulators and developers were introduced to World Bank EIA standards and practices in 1993, when the bank funded the Er Tan Hydro Station in Sichuan Province (Liu et al., 1993, 48–52). By the late 1990s, it had become standard for Chinese banks to require submission of both EIA and SIA reports – upon clearance by China's government agencies – before an investment loan could be granted. In 2004, the Chinese government ordered the suspension of the cascade development on the Nu River. In early 2013, development of the Nu River was revived as part of government planning (Li, 2013). This interval does indicate that EIA and SIA, aided by domestic and international NGO campaigns, were indeed playing a growing role in risk management for Chinese hydro companies (Brown and Xu, 2010; Magee and McDonald, 2009). Debate surrounding the Nu River Cascade centred on: (1) whether the EIA reports were publicly available and truthful; (2) how to balance development and nature conservation; (3) whether such big hydropower projects are beneficial to local people; (4) how to deal with forced migration; (5) whether electricity generated from big hydropower plants is green; and (6) whether it was wise to build dams in earthquake-prone areas (Li et al., 2012).

It is important to note that many Chinese hydropower policy researchers advocate basin-wide planning when considering applications for dam projects, especially when a development proposal involves dam cascades. Thus far, though, it seems that the only credible case of implementation in basin-wide planning was implemented on the Songliao River system in northeastern China from the 1980s to the mid-1990s (Wang and Wang, 1996). Meanwhile, there are numerous research initiatives that advocate improving China's record of sustainable hydropower development. For example, the China Institute of Water

Resource and Hydropower Research, first established in 1958 and administered by the Ministry of Water Resources, has, since 1994 had a Centre for Sustainable Hydropower Development.[9] Research by such research institutes, partly through international exchange, helps to raise the level of capacity in ecological and political/social risk management (Tang *et al.*, 2013).

To conclude, our study finds that among the different and increasingly competing interests associated with hydropower in China, the government's logic of economic growth and energy security can also accommodate arguments for ecological protection. An assortment of interests, domestic and international, is at play in China's hydropower industry. China has yet to find a well-functioning legal and bureaucratic framework for governing its hydro-energy industry. The idea of sustainable development is not foreign to Chinese debates about hydropower. The remaining question is how to have more of it.

Chinese involvement in hydropower development in the LMR

For the past several decades, the pursuit of hydropower development in the LMR has attracted support as well as criticism, both within the region and beyond. China features in the discussion for two major reasons. It insists on autonomously making plans to continue constructing mainstream dams on the upper Mekong River (the Lancang Jiang), i.e. showing little will to heed international calls for coordinated action in the entire basin (cf. Bakker, 1999). China also continues to partake in hydropower development projects in the LMR through aid, loans and project subcontracting (McDonald *et al.*, 2009).

An overall question then emerges: if the goal of Chinese policy towards the LMR is to pursue its acceptance as a neighbour, friendly to both the governments and peoples, what explains this pattern of Chinese behaviour? This section of the chapter offers an understanding of Chinese justifications and interests. It also offers an update on policy adjustments in response to calls for greater levels of environmental and corporate social responsibility.

The Greater Mekong Subregion (GMS) scheme as a political-diplomatic cover

Throughout most of contemporary history, China's economic ties with countries in the LMR were limited to maritime interactions (Glassman, 2010; Freeman, 2010). A change in central government policy came in 1992, when China's 'opening' policies, which had previously applied only to coastal regions, were applied to the provinces bordering China's neighbouring countries. Landlocked provinces, especially Yunnan, began to campaign for support from the central government to improve cross-border infrastructure as a means of linking China with economies in continental southeast Asia and further in south Asia as well (Summers, 2012).

Also in 1992, the Asian Development Bank (ADB) formally established an institutional structure for subregional economic cooperation between a group of countries linked together by the Mekong: Cambodia, Laos, Myanmar, Thailand and Vietnam. Prior to this, China was not a participant in regional institutions for LMR development. China was not invited to join the Mekong Committee from 1960 to 1995, mainly because it was not a member of the UN at the time, within which the Committee was couched. When the Mekong River Commission (MRC) was formed, China was invited to join but declined due to its fears over loss of sovereignty, and it only later became a dialogue partner. Within the GMS, Yunnan Province was chosen to represent the People's Republic of China (PRC), while Guangxi Province joined in 2005.

As Mr Li Fusheng of China Export-Import Bank stated

> The GMS framework is critical. Without GMS as a diplomatic cover, it would be unimaginable for a hydropower company from China to engage in either cross-border electricity connectivity projects or obtain central government approval for loans to a project in the LMR.'[10] China's diplomatic relationships with countries in the LMR continue to be complicated. For China, playing a role in the economic development of the LMR is integral to realizing its strategy of safeguarding its own borders, and electricity is considered as essential infrastructure.

Attraction from China's poverty-reduction approach

Another ideational justification for the continued Chinese involvement in energy development in the LMR is its poverty-reduction agenda. China has pursued phased development through a bottom–up approach, where local resources and village-level development and empowerment play an important role. The Chinese approach features:

> strong government commitment, active local participation, technological flexibility and diversity, a strong emphasis on rural development through agricultural and industrial activities and an emphasis on capacity building and training
>
> (Bhattacharyya and Ohiare, 2012, 676).

Accordingly, investment in hydropower development helps to improve the provision of electricity, which is essential for rural development and the associated urbanization process.

The continuation of extreme poverty at the household level, on top of the difficulties associated with rapid economic growth in most LMR countries, offers vast opportunities for Chinese engagement in their economic growth projects. Up until the 1990s, exporting raw materials in order to finance additional energy and resource extraction was a development strategy deployed by China. Through long-term 'coal/oil for machinery/finance' arrangements with Japan and other countries, China managed to overcome its own bottlenecks in energy supply (Ishikawa, 1987; Chow, 1992).

In other words, China perceives contemporary Laos, Cambodia and arguably Vietnam as occupying the same (economic) space as it did a few decades ago.

If China exported energy for the sake of overcoming poverty, then the LMR countries should do the same to yield similar poverty alleviation benefits. Particularly in areas immediately bordering its own, China seeks poverty alleviation in order to maintain social stability.

China also sees hydropower development in its foreign aid packages as part and parcel of the country's contribution to the United Nations' Millennium Development Goals (MDG) framework. For example, in the six-point pledge made by China at the 2008 UN High-Level Meeting on the MDGs, it was stated that 'In the coming five years, China will develop 100 small-scale clean energy projects for other developing countries, including small hydropower, solar power and bio-gas projects' (Xinhua News Agency, 2011).

In the eyes of some international observers, China's continued involvement in dam construction in the LMR is, diplomatically speaking, unwise. LMR societies have long complained that dams constructed by China upstream are a direct cause of droughts downstream (Pearse-Smith, 2012; McDonald *et al.*, 2009), although this has not been proven through any scientific endeavour. The drought in the summer of 2009 in the entire Mekong River system gave the MRC an impetus to convene. In April 2010, its first-ever summit took place, to foster a greater level of cooperation in the use of Mekong water resources. As an MRC dialogue partner, China was represented by its deputy foreign minister. This can be viewed as an indicator that China has more faith in formal government-to-government consultations than in reaching out to the societies at the grassroots level.

'Excess capacity' in the hydropower industry

China's hydropower development experienced a major slowdown in 2004, a year after the formal implementation of China's Environmental Impact Assessment Law, and the State Council's emergency measure to place on hold electric power projects that violated this law. Towards the end of the year, the State Council, which holds the final authority in approving large hydropower dams, decided to put on hold the implementation of the Memorandum of Understanding between the China Huadian Corporation and the Yunnan Provincial government, to start the first in a series of dams planned for the Nu River Cascade, which flows into Myanmar.

These measures resulted in 'excess capacity' in domestic hydropower expertise and equipment production. 'As a consequence, more and more hydropower corporations decided to seek operational opportunities abroad in order to survive.'[11] More investment in foreign projects, at the same time, had the potential to sustain the level of domestic capacity, which in turn could perpetuate investment abroad.

Adapting to international norms: green credit policy

In 2007, China's state-owned banks adopted a 'green credit' policy. The basic thrust of this policy is to hold Chinese companies accountable in environmental protection. Companies are required to submit EIA reports before their requests for loans are approved. Access to the dispersal of loans is tied to the monitoring of performance in environmental protection for the entire duration of a project. The key agencies for supervising the practice of this policy include the State Environmental Protection Administration, the China Banking Regulatory Commission and the People's Bank of China (Zhang *et al.*, 2011). In the same year, the policy was extended to cover loans to Chinese companies investing abroad, including two of its most active overseas development banks, the China Import-Export Bank (China Exim) and the Chinese Development Bank. It should be said that China's use of the banks as an agent to enforce environmental compliance by companies, internationally speaking, is fairly recent. Much still needs to be improved, both in the area of loan granting and in performance monitoring.

Since the mid-2000s, Chinese agencies, from government regulatory bodies to banks and stock exchanges, have issued over a dozen documents specifically requiring EIA and/or SIA reports by corporations (Bernasconi-Osterwalder *et al.*, 2012). In February 2013, China's Ministry of Commerce, the key agency for regulating FDI, and the Ministry of Environmental Protection, issued a new set of guidelines for environmental protection and corporate social responsibility (Ministry of Commerce and Ministry of Environmental Protection, 2013).

Through interviews with senior officers of the Export-Import Bank of China, including an active duty member of the bank's Project Appraisal Committee, we made several observations about the 'green credit' policy in practice. First, the policy has a limited scope of applicability; it applies only to those investments that require the Chinese party to contribute US$100 million or above. Investments below that figure do not require government approval. Second, assessment of the quality of preliminary EIA and SIA reports continues to be a challenge. In addition, it is seldom that a foreign government brings a case of alleged malpractice by a Chinese investor/operator to the bank for action. Third, the bank, borrowing a practice from the Ministry of Commerce for screening company eligibility for foreign aid projects, systematically weeds out those companies that are proven to have underperformed in their basic responsibilities. Last but not least, sources of financing for Chinese corporations investing abroad have diversified to such an extent that timely coordination between Chinese and international banking authorities is often costly and legally complicated, beginning with the identification of applicable jurisdictional power.[12]

China is still learning when it comes to its government-controlled banks being truly effective in enforcing government rules. We speculate that partly because the banks are themselves state-owned, there exists, in reality, a desire to walk the fine line between ensuring company profitability and taking them to task when it comes to environmental conservation infractions.

Does a Chinese hydropower corporation discipline itself?

Political risk – broadly meant here to include that growing out of social protest against the treatment of affected populations and the environment in the construction and operation of hydropower dams – is an inherent feature in cross-border investment. The institutional instruments that China employs to protect its corporations include risk insurance by its own banks, bilateral investment treaties (BITs) and membership in multilateral instruments designed to protect the free flow of investment among participating countries. Although China's central bank offered three political risk insurance services in 1983, their coverage was limited to expropriation, war and foreign exchange fluctuations. It was not until 2001 that China established a separate banking instrument – the China Export and Credit Insurance Corporation – to provide wider coverage for Chinese companies investing overseas. China has entered into BIT arrangements with all four countries in the LMR (with Thailand in 1985, with Vietnam in 1992, with Laos in 1993 and with Cambodia in 1996). The Multilateral Investment Guarantee Agency (MIGA) is an important multilateral instrument for managing cross-border investment risks. China signed and approved the MIGA convention in 1988.

The extent to which Chinese hydropower corporations make use of any category of these instruments as a tool for managing their exposure to political risk in the LMR is difficult to ascertain. Our interviews with China's large hydropower companies[13] informed us that companies find it necessary to employ a combination of tools. These include: (1) diversification of capital structure in project financing, through the incorporation of capital from investors in the host country; (2) placing large, long-term projects under the government-to-government economic cooperation framework; (3) enlisting local government(s) at project sites as cooperation partner(s); (4) developing amicable relationships with the local communities in and close to project sites; and (5) participating in MIGA and China's own investment insurance schemes.

Chinese hydropower company executives are of course aware of the complaints and protests regarding their operations in LMR projects. One experienced CEO offered the following insights. First, there is broad awareness that models for managing project–community relationships in China are not suitable for Chinese investment in the LMR countries. The key difference lies in the government's role. In China, as stated before, it is the government that decides the terms of compensation for resettlement and livelihood restorations plans and strategies. In the LMR countries, however, central and local government tend not to be as interventionist. This frequently leaves the Chinese corporation unprepared to come up with timely and locally acceptable solutions.

Second, Chinese companies, especially during the preparation and dam construction stages of a project, often have to meet demands from local communities that are not within the scope of the contract. The grounds for making such demands are usually based on the fact that the central government has failed to sufficiently consult with local communities for the terms of compensation.

Related to this is a third factor: the margin of profit in contract design. Because Chinese investors have to compete against each other over the same project offer, the winner usually emerges by making large concessions. As a result, they prefer to donate schools, the construction of drinking water facilities for the villagers, and donations for temple repairs. These incur costs that are easier to control than having to satisfy ad hoc demands, which may increase.

In our research, we raised the question of standards and norms for addressing tasks like compensation and environmental protection. We learned that Chinese companies prefer to observe the rules and standards set by the host country governments. When such rules and standards are non-existent, companies choose to go through a negotiation process with whoever approaches them. There has yet to be a strong interest in the application of those standards used in China as a remedy.

Concluding observations

Proceeding from the earlier discussions in this chapter, we can make several general observations about China's pursuit of hydropower, both within China and the LMR. First, Chinese involvement – by way of development aid or commercial bidding – is driven by structural interests that are ideational, institutional, and industry-specific. Second, the Chinese hydropower industry has a history of socializing with international governmental and industry programmes and efforts aimed at pursuing sustainable development hydropower within the country. Norms such as environmental impact assessment and social impact assessment are well institutionalized in the formal processes of Chinese hydropower development. Third, the key issue that divides Chinese industry perspectives on hydropower and those outside China seems to be what constitutes an appropriate division of responsibility when it comes to protecting the rights and interests of the affected populations and the environment.

Since an important goal of scrutinizing the political ecology of China's hydropower industry is to help improve the level of acceptability of Chinese investment projects in the LMR, especially those by the state-owned corporations, we find it worthwhile to suggest that international observers pay particular attention to the following, though far from exhaustive, issues. The goal is to foster greater levels of awareness among Chinese actors in the LMR to adopt a multiple stakeholder approach.

First, there is a need to focus on Chinese hydropower company executives who operate in the LMR. They need to fully appreciate the role that stakeholder communities play in hydropower projects. Increased empathy and a greater understanding of the issues among those executives can make a significant difference in everyday decision-making, especially concerning how impacts are mitigated. For the community, the tradition of placing a great level of faith in the government – its approval in project entry and intervention when societal dissatisfaction arises – though partly rooted in domestic experiences, continues to be strong. In the LMR context, the community of stakeholders goes far beyond the geographical region. There arises the challenge of mutually reaching out, with the aim of seeing changes in behaviour towards the mitigation of future deficiencies in managing both the human and environmental impacts associated with project development.

Second, efforts should be made to promote the harmonizing of industry standards for the appropriate care of the affected populations and the environment. Thus far, partly because project bidding is a competitive process for all of the formal entities involved, the general approach has been to promote 'best practice' cases and hope that these models will imitated by other projects. But such changes do not automatically take place. Therefore, a task for the stakeholder community is to bring hydropower developers, governmental and non-governmental actors, and the affected populations together and develop project-site specific rules and standards.

In a nutshell, the real challenge lies in the detail when advocating improvements in hydropower development, both in China and in the LMR. 'Learning by doing' is a natural approach when it comes to FDI management. Our observation is that the Chinese hydropower industry has demonstrated a positive propensity for learning, including from international sources.

Notes

1 We thank our interviewees in Beijing, among them Professor Chen Wenying of the Department of Hydrological Engineering, Tsinghua University, for pointing this out to us.

2 External funding consisted of a loan from the World Bank, and grants from the governments of Norway and Australia.

3 Interview with a senior researcher, Development Research Centre, Ministry of Water Resources, Beijing, 15 Sept. 2012.

4 Development Research Center, Ministry of Water Resources, www.waterinfo.com. cn/English.

5 Chinese Hydrological Society, www.ches.org.cn/zgslxh/english.

6 Interview with a senior researcher in the Development Research Center, Ministry of Water Resources, Beijing, 15 Sept. 2012.

7 Interview note, ERI, Beijing, 11 May 2013.

8 Professor Shi Guoqing, Director of National Research Center for Resettlement, Hohai University, Nanjing, China. Personal communication with the author, 20 Dec. 2012.

9 Center for Sustainable Hydropower Development, China Institute of Water Resource and Hydropower Research, www.hydro.iwhr.com.

10 Note from the workshop on China and LMR hydropower development, organized by the author, 16 Dec. 2012. Quoting Li Fusheng, China Export-Import Bank.

11 Notes from the workshop on China and LMR hydropower development, organized by the author, 16 Dec. 2012. Quoting Wu Yusong, WWF Yunnan office.

12 Author's follow-up interview with Li Fusheng, of the Export-Import Bank, 24 March 2013, Beijing.

13 Our key sources of input are from Huaneng Lancang River Hydropower Company and Sinohydro.

References

Bakker, K. (1999) 'The politics of hydropower: developing the Mekong', *Political Geography*, 18(2): 209–32.

Bernasconi-Osterwalder, N., Johnson, L., and Zhang, J. (2012) *Chinese Outward Investment: An Emerging Policy Framework*, Winnipeg: International Institute for Sustainable Development, pp. 83–9.

Bhattacharyya, S. C., and Ohiare, S. (2012) 'The Chinese electricity access model for rural electrification: Approach, experience and lessons for others', *Energy Policy*, 49: 676–87.

Blackman, A., and Wu, X. (1999) 'Foreign direct investment in China's power sector', *Energy Policy*, 27: 695–711.

Brown, P. H., and Xu, K. (2010) 'Hydropower development and resettlement policy on China's Nu River', *Journal of Contemporary China*, 19(66): 777–97.

Bryant, R. L., and Bailey, S. (1997), *Third World Political Ecology*, London: Routledge.

Chen, J. (1998) 'The development and conception of administrative law in the PRC', *Law in Context*, 16(2): 72–105.

Cheng, Y., Wong, W., and Woo, C. (2013) 'How much have electricity shortages hampered China's GDP growth?', *Energy Policy*, 55: 369–73.

China Export–Import Bank (2007) 'Guidelines for Environmental and Social Impact Assessments of the China Export and Import Bank's (China Exim Bank) Loan Projects', revised in 2007, unofficial translation retrieved from www.globalwitness. org/sites/default/files/library/Chinese%20guidelines%20EN.pdf, accessed July 2013.

Chow, L. C. H. (1992) 'The changing role of oil in Chinese exports, 1974–1989', *China Quarterly*, 131: 750–65.

Compilation Board (2012) *Laws and Rules for Water Works and Hydropower*, Beijing: China Water Publishing House.

Conrad, B. (2012) 'China in Copenhagen: Reconciling the "Beijing climate revolution" and the "Copenhagen climate obstinacy"', *China Quarterly*, 210: 435–55.

Dai, Q. (1998) *The River Dragon has Come! The Three Gorges Dam and the Fate of China's Yangtze River and its People*, New York: M. E. Sharpe.

Economy, E. C. (2002) 'China's go west campaign: Ecological construction or ecological exploitation', *China Environment Series*, 5: 1–12.

Economy, E. C. (2005) *The River Runs Black: The Environmental Challenge to China's Future*, Ithaca, NY: Cornell University Press.

Freeman, C. (2010) 'Fragile edges between security and insecurity: China's border regions', in R. Guo and C. Freeman (eds), *Managing Fragile Regions*, Berlin: Springer, pp. 23–46.

Geheb, K., and Mapedza, E. (2008) 'Political ecologies in natural resources management', in D. Bossio and K. Geheb (eds), *Conserving Land, Protecting Water: 'Bright Spots' Reversing the Trends in Land and Water Degradation*, Wallingford: CAB International, pp. 51–68.

Glassman, J. (2010) *Bounding the Mekong: The Asian Development Bank, China, and Thailand*, Honolulu, HI: University of Hawaii Press.

Heggelund, G. (2004) *Environment and Resettlement Politics in China: The Three Gorges Project*, Farnham: Ashgate Publishing.

Hu, X. (2012) 'There is no "dam demolition movement" in the United States', *China Energy News*, 15 Oct., p. 20; in Mandarin: 胡学萃, '美国不存在'拆坝运动,' 《中国能源报》, 2012 年10月15日, 第20版

Huang, H., and Yan, Z. (2009) 'Present situation and future prospect of hydropower in China', *Renewable and Sustainable Energy Reviews*, 13(6–7): 1652–6.

IRENA (International Renewable Energy Agency) (2012) *Renewable Energy Technologies: Cost Analysis Series*, 1(3–5), Abu Dhabi: IRENA.

Ishikawa, S. (1987) 'Sino-Japanese economic cooperation', *China Quarterly*, 109: 1–21.

Li, J. (2013) 'Ban lifted on controversial Nu River dam projects', *South China Morning Post*, 15 Jan., www.scmp.com/news/china/article/1135463/ban-lifted-controversial-nu-river-dam-projects, accessed Jan. 2014.

Li, W., Liu, J., and Li, D. (2012) 'Getting their voices heard: Three cases of public participation in environmental protection in China', *Journal of Environmental Management*, 98: 65–72.

Lin, C. (2005) 'Observation and consideration of anti-dam movement and dam demolition in the United States', *China Three Gorges Construction*, Z1: 44–57; in Mandarin: 林初学，'美国反坝运动及拆坝情况的考察和思，'《中国三峡建设》，2005年Z1期，第44–57页。

Liu, Z., Yu, W., Shi, P., and Duan, K. (1993) 'Insights from the World Bank's environmental impact assessment on the Er Tan Hydropower Station', *Sichuan Environment*, 1: 48–52; in Mandarin: 刘珍海、喻卫奇、石澍滋、段开甲，'从世界银行对二滩水电站环境影响评估中得到的启示，'《四川环境》1993年01期，第48–52页

Lubman, S. B. (2012) *The Evolution of Law Reform in China: An Uncertain Path*, Northampton, MA: Edward Elgar.

Luo, Y. Y. (2007) 'Our country's hydropower development financing', *Reports on Financing in China*, 2007 edn, pp. 65–123, archived at www.cnki.net, accessed Jan. 2014.

Ma, D., Li, Z., and Wang, X. (2006) 'Problems in utilization of foreign capital in hydropower and water works projects and remedies', *Water Resources and Hydropower of Northeast China*, 4: 60–70.

Ma, J. (2011) 'On-grid electricity tariffs in China: Development, reform and prospects', *Energy Policy*, 39(5): 2633–45.

Magee, Darrin and McDonald, Kristen. (2009). 'Beyond Three Gorges: Nu River hydropower and energy decision politics in China', *Asian Geographer* 25(1–2): 39–60.

Martinot, E. (2001) 'World Bank energy projects in China: influences on environmental protection', *Energy Policy*, 29(8): 581–94.

Matthews, N. and Schmidt, J. (2014) 'False promises: the contours, contexts and contestation of good water governance in Lao PDR and Alberta, Canada', *International Journal of Water Governance* 2(1): 1–20.

McCormack, G. (2001) 'Water margins: Competing paradigms in China', *Critical Asian Studies*, 33(1): 5–30.

McDonald, K., Bosshard, P., and Brewer, N. (2009) 'Exporting dams: China's hydropower industry goes global', *Journal of Environmental Management*, 90(3): S294–S302.

McGregor, J. (1993) 'Going it alone: Beijing is pushing ahead on the Three Gorges project', *Asian Wall Street Journal*, 20 Jan., p. 1.

Mertha, A. C. (2008), *China's Water Warriors: Citizen Action and Policy Change*, Ithaca, NY: Cornell University Press.

Mertha, A. C., and Lowry, W. R. (2006) 'Unbuilt dams: Seminal events and policy change in China, Australia, and the United States', *Comparative Politics*, 39(1): 1–20.

Middleton, C., Garcia, J., and Foran, T. (2010) 'Old and new hydropower players in the Mekong Region: Agendas and strategies', in F. Molle, T. Foran, and M. Käkönen (eds), *Contested Waterscapes in the Mekong Region: Hydropower, Livelihoods and Governance*, Singapore: Institute of Southeast Asian Studies, pp. 23–54.

Ministry of Commerce and Ministry of Environmental Protection, China (2013) *Guidelines for Environmental Protection in Foreign Investment and Cooperation*, archived at http://english.mofcom.gov.cn/article/policyrelease/bbb/201303(20130300043226).shtml, accessed May 2014.

Morton, K. (2007) 'Transnational advocacy at the grassroots: Benefits and risks of international cooperation', in P. Ho and R. L. Edmonds (eds), *China's Embedded Activism: Opportunities and Constraints of a Social Movement*, New York: Routledge, pp. 195–215.

Pearse-Smith, S. (2012) '"Water war" in the Mekong Basin?', *Asia Pacific Viewpoint,* 53(2): 147–62.

PRC (People's Republic of China) (2004) 'Water Law of the People's Republic of China', archived at: www.china.org.cn/english/government/207454.htm, accessed July 2013.

Shan, B., Xu, M., Zhu, F., and Zhang, C. (2012) 'China's energy demand scenario analysis in 2030', *Energy Procedia*, 14: 1292–8.

Shyu, C. W. (2012) 'Rural electrification program with renewable energy sources: An analysis of China's township electrification program', *Energy Policy,* 51: 842–53.

Sovacool, B. K. (2008) 'The costs of failure: A preliminary assessment of major energy accidents, 1907–2007', *Energy Policy,* 36: 1802–20.

Summers, T. (2012) '(Re)positioning Yunnan: Region and nation in contemporary provincial narratives', *Journal of Contemporary China*, 21(75): 445–59.

Tan, Y., and Yao, F. (2006) 'Three Gorges Project: Effects of resettlement on the environment in the reservoir area and countermeasures', *Population and Environment*, 27(4): 351–71.

Tang, W., Li, Z., Qiang, M., Wang, S., and Lu, Y. (2013) 'Risk management of hydropower in China', *Energy*, 60(1): 316–24.

Tilt, B. (2007) 'The political ecology of pollution enforcement in China: A case from Sichuan's rural industrial sector', *China Quarterly*, 192: 915–32.

Tortajada, C., Altinbilek, D., and Biswas, A. K. (2012) *Impacts of Large Dams: A Global Assessment,* New York: Springer.

Wang, J., and Wang, D. (1996) 'The planning of water resources development on large rivers in northeastern China', *Journal of Hydroelectric Engineering,* 3: 1–13; in Mandarin: 王羿野，王丹予,东北几大江河水利水电开发规划总结，《水力发电学报》，1996年第3期，第1–13页.

Wang, Y., Morgan, R. K., and Cashmore, M. (2003) 'Environmental impact assessment of projects in the People's Republic of China: New law, old problems', *Environmental Impact Assessment Review,* 23: 543–79.

Wang, Z. Y., Gao, H., and Zhou, D. (2006) 'China's achievements in expanding electricity access for the poor', *Energy for Sustainable Development,* 10(3): 5–16.

Wang, P., Wolf, S. A., Lassoie, J. P., and Shikui, D. (2013) 'Compensation policy for displacement caused by dam construction in China: An institutional analysis', *Geoforum,* 48: 1–9.

Wouters, P., Hu, D., Zhang, J., Dan Tarlock, A., and Andrews-Speed, P. (2004) 'The new development of water law in China', *University of Denver Water Law Review,* 7: 243–308.

Xi, J., and Hwang, S. (2011) 'Relocation stress, coping, and sense of control among resettlers resulting from China's Three Gorges Dam Project', *Social Indicators Research,* 104(3): 507–22.

Xinhua News Agency (2011) 'Six measures for foreign aid pledged by the Chinese Government at the 2008 UN High-Level Meeting on the Millennium Development

Goals', retrieved from http://news.xinhuanet.com/english2010/china/2011-04(21)/c_13839683_20.htm, accessed July 2013.

Xiong, X., Zhao, M., and Kang, Y. (2010) 'Current status and perspective of hydropower CDM project development in China', *China Water Power and Electrification*, 6: 1–7.

Xu, J., and Melick, D. R. (2007) 'Rethinking the effectiveness of public protected areas in southwestern China', *Conservation Biology*, 21(2): 318–28.

Yang, M. (2003) 'China's rural electrification and poverty reduction', *Energy Policy*, 31(3): 283–95.

Yang, X. (2004) 'A brief analysis of dam removal in the United States', *China Water Resources*, 14: 15–20; in Mandarin: 杨小庆，'美国拆坝情况简析，'《中国水利》, 2004年13期 15–20

Yong, M., and Grundy-Warr, C. (2012) 'Tangled nets of discourse and turbines of development: Lower Mekong Mainstream dam debate', *Third World Quarterly*, 33(6): 1037–58.

Zeng, M., Xue, S., Li, L., and Wang, Y. (2013) 'China's large-scale power shortages of 2004 and 2011 after the electricity market reforms of 2002: Explanations and differences', *Energy Policy*, 61: 610–18.

Zhang, B., Yang, Y., and Bi, J. (2011) 'Tracking the implementation of green credit policy in China: Top–down perspective and bottom–up reform', *Journal of Environmental Management*, 92(4): 1321–7.

Zhao, X., Liu, L., Liu, X., Wang, J., and Liu, P. (2012) 'A critical analysis of the development of China's hydropower', *Renewable Energy*, 44: 1–6.

Zhou, X., Yi, J., Song, R., Yang, X., Li, Y., and Tang, H. (2010) 'An overview of power transmission systems in China', *Energy*, 35: 4302–12.

4 From Manwan to Nuozhadu

The political ecology of hydropower on China's Lancang River

Xing Lyu

Introduction

This chapter examines the drivers and institutional arrangement of hydropower development projects on the Lancang River – as the upper Mekong River is known in China. Hydropower development projects can have significant impacts on river flows, the environment and the livelihoods of people. Hence, the World Commission of Dams (WCD) (2000) argued that improving decision-making is key to addressing these impacts. Decision-making is highly complex, however. This complexity results not only from the dynamic drivers that push hydropower projects forward but also from the changing institutional arrangement of state, market and society. The decision-making on the hydropower development on the Lancang is very illustrative. Hydropower development is occurring at a time when China is experiencing rapid economic change and the transition from a planned to a market economy. Dynamic drivers and a changing institutional landscape are shaping the course of hydropower development on the Lancang. This chapter aims to present Chinese perspectives of hydropower development specifically on the Lancang mainstream, drawn from data from the author's field studies and the Chinese literature.

In China, increasing electricity demands and an uneven distribution of energy sources has created opportunities to generate electricity in one place for transmission to another. Rapid economic development and improving living standards have increased demand and driven the expansion of electricity production. Annual electricity production increased nearly fourteen-fold, from 300.63 billion kWh in 1980 to 4207.16 billion kWh in 2010. Hydropower also grew by 12.4 times, from 58.21 billion kWh to 722.17 billion kWh during the same period. In 2010, coal contributed 79.20 per cent of this electricity production, hydropower 17.12 per cent and other sources the remainder (China Statistics Bureau, 1996, 2012).

Most of the nation's coal is found in northern China and hydropower in its southwest, while the major electricity markets are in the eastern and southern provinces of China. Yunnan Province is energy-rich in terms of hydropower. Its technical hydropower potential accounts for 17.9 per cent of China's total (Ma, 2003). The Lancang is regarded as 'fu kuang' or rich ore with its hydropower potential of 36,560 MW (Ma, 2004). The Lancang originates on the Qinghai Plateau, passes eastern 'xizhan' or Tibet, then enters Yunnan, and, once it crosses the border with Laos, it continues southwards as the Mekong. The Lancang runs for 2,180 km with a drop of 4,800 m. Most of the Lancang valley, especially in Yunnan, is a 'V' shape, which is ideal for hydropower development. Fifteen mainstream dams have been planned on the Lancang. In 1986, construction of the first mainstream dam – the Manwan – started, and by 2014 six of 15 planned projects were in operation. These six mainstream projects are (from south to north): the Jinghong (1750 MW), the Nuozhadu (5850 MW), the Dachaoshan (1350 MW), the Manwan (1550 MW), the Xiaowan (4200 MW) and the Gongguoqiao (900 MW). The development of these mainstream hydropower projects spanned a period of 26 years and manifests the dynamic drivers and changing institutional framework that have evolved since the construction of the Manwan through to the Nuozhadu, the most recently commissioned project.

This chapter points out that the state firmly controls the development of large and medium hydropower projects through various legal, administrative and management instruments. The administration and regulatory functions are shared among state agencies at the central and provincial levels. The management functions are granted to state-owned enterprises (SOEs) – legal entities answerable to state-dominated shareholders, which implement state-led development strategies. SOEs have dominated the energy sector at both the central and provincial levels. Consulting companies, for example, HydroChina Kunming Engineering Corporation and China International Engineering Consulting Corporation, play crucial roles in providing technical services both to the government and to developers. Pro-hydropower scholars and developers can also assist in legitimizing and resolving issues of hydropower. Environmental journalists and civil society activists criticize the lack of caution in hydropower development, while critical scholars prudently draw attention to its negative impacts. Affected communities are excluded from the debate and major decision-making processes, and become objects rather than subjects of the discussion between the state and civil society in the public domain. Access to water resources or the development of mega water projects is mainly regulated through legal, market and political processes. To improve hydropower decision-making, there is a need to strengthen the cumulative impact assessment, transparency and open, respectful and constructive deliberation surrounding major decisions.

The remainder of this chapter is divided into four parts. The first part reviews the trajectory of hydropower development on the Lancang mainstream. It presents the important points in history when actors have pushed forward their respective agendas. The second part will analyse the rhetoric of key actors, including the dominant rhetoric of developers, consulting companies, the government and pro-hydropower scholars, and the alternative rhetoric of several environment journalists, civil society activists and scholars. The third part will discuss the social and environmental impacts of hydropower development. Finally, the fourth part will analyse the institutional factors and power relations between the actors and draw some conclusions.

Trajectory

This section will retrospectively review the trajectory of hydropower development on the Lancang mainstream and the key events, as discussed by Lin (1993) and Ding (2005). Many actors and organizations are engaged in the development of hydropower on the Lancang (Table 4.1). They include government agencies, developers, consulting companies and investors. State-owned banks are not included in Table 4.1 because they are not directly engaged in the decision-making process.

Milestones to Manwan

Early development initiatives involved a complex mix of government actors and industry, aiming to assess the technical feasibility of development. During this period, the main actors included the Yunnan Provincial Government (YPG), consulting companies and water resources agencies.

Studies on hydropower development on the Lancang were initiated as early as 1956. In 1956, the HydroChina Kunming Engineering Corporation (HCKEC), as it is now known, started preparing a water resources investigation of the Lancang. The following year, it undertook a nine-month field inventory to assess water resources availability. The study investigated 1,148 km of Lancang watercourses in Yunnan, and identified 21 potential hydropower project sites. In 1958, the HCKEC also conducted primary planning for the Gongguoqiao–Xiaowan section and a detailed survey of the Xiaowan site. Meanwhile, the YPG invited the chair of the Yellow River Conservancy Commission to carry out a study tour of Yunnan's major rivers. As a result of the tour, the Liutongjing Hydrologic Station was established on the Lancang. All of these study activities were suspended in 1959, at the start of the Great Leap Forward and the Cultural Revolution (1959 to 1976), but they resumed in 1969, when the YPG requested the HCKEC to identify an energy source that could supply electricity for iron ore mining and smelting in Yunnan's Xinping County. The HCKEC then conducted a field inventory of the middle section of the Lancang between the tributaries of the Yangbi and Tuoge Rivers. Manwan was proposed after a comparison of three hydropower sites, at Manwan, Sijia Village and Baga, in April 1970. Before Manwan could be approved, however, a hydropower development plan was required in order to optimize the use of the river.

Hydropower development plan

The Lancang River hydropower development plan is an essential document that details technical conclusions and proposes future decision-making actions. In 1973, the Ministry of Water Resources and the Power Industry (MoWRPI) instructed the HCKEC to undertake planning for the Middle Lancang (between Gongguoqiao and Manwan). The Planning Report on the Development of the Middle Lancang River was submitted in 1977. Manwan was again

Table 4.1 Major actors and their roles

Actor	Acronym	Role	Client
National Development and Reform Commission	NDRC	Regulator	Public
Ministry of Water Resources (since 1988)	MoWR	Regulator	Public
Ministry of Water Resources and Power Industry (1958–1979 and 1982–1988)	MoWRPI		
Ministry of Power Industry (1955–1958, 1979–1982, 1993–1998)	MoPI		
Ministry of Energy (1988–1993)	MoE		
Ministry of Environment Protection	MoEP	Regulator	Public
Minstry of Land and Resources	MoLR	Regulator	Public
Yunnan Provincial Government	YPG	Regulator	Public
Huaneng Lancang River Hydropower Corporation	HLRC	Developer	Shareholder
Yunnan Dachaoshan Hydropower Corporation	YDHC	Developer	Shareholder
HydroChina Kunming Engineering Corporation	HCKEC	Consulting	Developer
Hydropower Planning & Design General Institute	WRHPDGI	Consulting	MoWR
China International Engineering Consulting Corporation	CIECC	Consulting	NDRC
State Power Corporation of China (1997–2002)	SPCC	Investor	State Council
State Energy Resource Investment Corporation	SERIC	Investor	State Council
State Development Investment Corporation	SDIC	Investor	State Council
Yunnan Provincial Investment Corporation	YPIC	Investor	YPG
Hongtao Group		Investor	Shareholders
Guangdong Provincial Government	GPG	Electricity buyer	Public
China Southern Power Grid Corporation	CSG	Electricity transmission	Electricity users

Source: compiled by the author.

recommended in the report. From 1978 to 1980, the HCKEC continued its inventory survey on the Upper Lancang (upstream of Gongguoxiao) and its study on the development of a hydropower cascade in the Upper and Middle Lancang. In December 1980, a group of experts in mining, hydropower and transportation from Belgium were invited by the central government to visit the Middle Lancang. They shared their knowledge and experiences of river development and supported the idea of constructing the Manwan Dam. This was first due to its geographic location – it was a more accessible project site at the time – and also because of the detailed technical studies available.

In 1979, the Ministry of Power Industry (MoPI) officially declared the Lancang River as one of ten key energy development bases. In 1980, the HCKEC submitted the Report on the Manwan Dam Selection and the Draft Planning Report on the Development of the Middle Lancang River. In the Planning Report it also recommended a sequence of cascade dams, with the Xiaowan as the tallest dam in the Middle Lancang. In late 1980, Mr Luo from the MoPI claimed to have completed planning for the Middle Lancang and proposed that planning for the Lower Lancang River (between Manwan and the national border) be commenced in order to finalize the cascade hydropower projects.

In 1981, the HCKEC submitted to the MoPI the Inventory Survey Report on the Lower Lancang River. Nuozhadu was identified as the best candidate for another tall dam. The HCKEC also completed the feasibility study for the Manwan Dam. In 1982, the Water Power Inventory and Design Institute of the MoPI organized a reporting workshop on the Manwan Feasibility Study in Beijing. In 1983, the Water Resources and Hydropower Planning & Design General Institute of the MoWRPI held a technical workshop on the site selection for the Manwan Dam in Kunming, followed by a reporting workshop on the Manwan Dam type selection in Beijing. In 1986, the planning for the Middle–Lower Lancang development was completed, which integrated the Middle Lancang Plan completed in 1980.

The MoWRPI, the MoPI and the YPG, with the assistance of consulting companies, were actively engaged in the development planning of the Lancang River. Technical documents were a component of the development process, but the real decision-making for these mega-projects took place in the political arena.

Integration of Manwan into the National Five-Year Development Plan

In order to obtain both administrative approval and budget support from the central government, the YPG and the MoWRPI had to convince the National Development and Reform Commission responsible for coordinating and channelling financial resources according to the National Development Plan. Any mega-project must first obtain administrative approval for the initial proposal and feasibility study, and starting and completion construction approvals from the National Development and Reform Commission. A mega-project must be integrated into the National Five-Year Plan and Annual Action Plan before it can be launched. During the time of the Manwan Project development, financing from capital markets was very limited. It was impossible to implement a mega-project without financial support from the central government. The NDRC's approval signified that a project had secured its financial support as well as administrative permission. The YPG and the MoWRPI thus began to lobby hard for approval of the Manwan Project.

In 1984, the YPG formally appealed to the State Council to include Manwan in the Seventh National Five-Year Plan. The 70 Yunnan delegates also appealed to the Second Session of the Sixth National People's Congress, another arena that could lead to approval. Pu Chaozhu, YPG governor, paid a visit to Qian Zhengying, Minister of MoWRPI, and proposed that the Manwan be jointly funded by both the YPG and MoWRPI. He proposed that Yunnan would contribute RMB 300 million, which was estimated to account for 30 per cent of the construction costs at that time. This idea of joint funding was novel to Chinese hydropower decision-making, and the minister and the governor agreed that it seemed to be the way forward. The YPG then detailed its share of the construction costs to include a lump sum payment for the resettlement plan implementation and construction materials. In order to obtain the approval, the YPG also proposed adopting a bidding scheme for the dam construction – a very pioneering suggestion at that time – and to establish a YPG-MoWRPI joint steering committee responsible for the Manwan Dam (Liu and Chen, 1992).

In July 1984, a group of experts from the Chinese Academy of Sciences was requested by the National Development and Reform Commission to undertake a study on issues of efficient resources exploitation and production layout. The resultant study contained a policy brief that urged the government to make the Manwan development a priority. In the second half of 1984, Yue Shihua, the Director of Yunnan Provincial Electricity Department, on behalf of the YPG, paid a visit to Beijing and briefed relevant ministries on the Manwan project's financial arrangements. The MoWRPI then submitted a report to the National Development and Reform Commission requesting inclusion of Manwan in the Seventh Five-Year Plan. Minister Qian himself wrote to Song Renping, Chairperson of the National Development and Reform Commission, supporting this request. Meanwhile, the MoWRPI and the YPG jointly reviewed the draft design of the Manwan project and then recommended again the inclusion of Manwan in the Seventh Five-Year Plan. Song was very positive about the report and recommendation (Liu and Chen, 1992).

In support of the integration of Manwan into the 1985 Annual National Development Plan, the MoWRPI formally approved the draft design of the Manwan Project. The YPG and the MoWRPI respectively submitted reports requesting the inclusion of Manwan into the 1985 Action Plan. The YPG and the MoWRPI then established 'the Leading Group for Manwan's Construction' and an 'Advisory Group for Manwan', built around their joint funding agreement. Finally, a preparatory Manwan Management Bureau was established.

On 12 April 1985, the National Development and Reform Commission approved the integration of Manwan into the 1985 Annual National Development Plan and gave the green light to prepare for construction. The MoWRPI immediately agreed that the Manwan Management Bureau should be responsible for its construction as an executive agency. The YPG formed the YPG's Office for Assisting the Manwan Construction. After 19 years of preparation, construction of the Manwan commenced in 1985. In July 1993 the first generator began to produce electricity and in June 1995 the other five generators began operation. The second phase of the Manwan project, an underground powerhouse that increased the total generation capacity from 1250 MW to the present 1500 MW, was completed in 2007.

The YPG was very active in promoting Manwan as a major contribution to economic development in Yunnan. Based on hydropower and mineral potential, the YPG formulated a development strategy titled, 'dianli youxian kuangdan jiehe', or 'Advancing Electricity First and Combining Development of Mining and Electricity Industries'. The Manwan Dam has a potential annual generation capacity of 7.8 billion kWh, which is valued at more than RMB 20 billion a year (approx. US$3.2 billion), equivalent to Yunnan's entire industrial output in 1984 (Liu and Chen, 1992). The HCKEC prepared the necessary technical documents that were subject to approval by the MoWRPI at the central level. The Water Resources and Hydropower Planning & Design General Institute provided ministries with the technical services necessary to review these documents. As noted above, Manwan had to be integrated into the national development plans, and the National Development and Reform Commission was in charge of this integration. Prior to Manwan, the central government had been the sole investor in this kind of mega-project. Manwan was the first project to be jointly financed by the central and provincial governments, giving rise to the so-called 'Manwan Model'. This joint financing was a strategic choice by the YPG to appeal to the central government by demonstrating local buy-in to the project. This institutional arrangement under the planned economy gradually gave way to a new institutional arrangement under the market economy.

Forward to Nuozhadu

Immediately after construction of the Manwan Dam commenced, the YPG continued to persuade both the central government and other investors to secure administrative permission and financial support for the Dachaoshan Project. The energy sector reforms from the mid-1980s to the early 2000s dramatically changed China's hydropower landscape, however (see Chapter 3 in this volume). As a result, SOEs began to take the leading role in promoting the Lancang hydropower cascade. The first reform was initiated in 1985 under the principle of 'zhengqi fengkai', the separation of administration from management, and this redefined the role of the government and public corporations. The MoPI was transformed into the State Power Corporation of China (SPCC) between 1996 and 1998. The second reform, implemented in 2002, split the SPCC into 11 corporations (Magee, 2006; Dore *et al.*, 2007). The China Huaneng Group (CHG) is one of these, and it has, since then, played a profound role in the hydropower development of the Lancang mainstream through a subsidiary.

New electricity market and financial sources

The success of Manwan encouraged the YPG to seek support from the central government and new financial sources as well as from the electricity market. Both the YPG and hydropower industry recognized that if the Lancang Hydropower Development Plan was to be implemented in full, new markets outside Yunnan would have to be identified for the surplus electricity. In 1985, the HCKEC conducted a study on the transmission infrastructure between Yunnan Province and Eastern China. The study projected the supply and demand for electricity in Sichuan, Yunnan, Guizhou, Guangxi, Guangdong and Hainan provinces, and concluded that Guangdong would be the province with the greatest electricity deficit between 1995 and 2010. Only Yunnan and Guizhou would have surplus electricity (Qing, 1997). In 1988, the Agreement on the Joint Exploitation and Transmission of Yunnan's Electricity to Guangdong Province was signed by the Ministry of Energy, the State Energy Resources Investment Corporation, the YPG and the Guangdong Provincial Government. Yunnan then proposed and negotiated the transmission of electricity with the support of the central government. Later, this initiative was given the name 'xidian dongsong' or West-East Electricity Transmission, and was integrated into the 'xibu dakaifa' China Western Development Strategy.

The YPG then submitted an appeal to the State Council to adopt a new business model to develop water resources on the Middle and Lower Lancang River. The appeal proposed the establishment of a hydropower development company that would be responsible for hydropower development on the Lancang. In April 1991, in Yunnan's Jinghong, the Ministry of Energy, State Energy Investment Corporation, the Guangdong Provincial Government and the YPG signed the Agreement on the Seasonal Transmission of Electricity from Yunnan to Guangdong, and the Agreement on the Joint Development of Cascade Hydropower Projects on the Middle–Lower Lancang Mainstream.

The YPG also mobilized scholars to support the Lancang hydropower development and influence the central government. In 1988, a group of outstanding scholars was invited to the Lancang River to conduct a study tour and it concluded that hydropower development should be prioritized because it would have spillover effects on other sectors such as mining and the processing industry (Yu *et al.*, 1988). Their study report was submitted to the central government. On 26 July 1988, China United Associations of Economic Research organized a workshop on Strategies of Integrated Economic Development of the Lancang River Basin in Beijing, which was attended by more than 300 participants comprising scholars and officials. At the workshop, the YPG's governor delivered a speech in which he stressed that 'Yunnan will advance its economy when the development of the Lancang's hydropower reaches its full potential' (Gao, 1988). In 1989, another workshop was organized to analyse cascade hydropower projects and it was suggested that, following Manwan, the construction sequence should be Dachaoshan, Xiaowan and then Nuozhadu.

Dachaoshan was the second mainstream project. The YPG successfully secured investment from the State Development Investment Corporation. The Yunnan Dachaoshan Hydropower Corporation was registered as a shareholder company in 1994 with capital of RMB 1.77 billion. The State Development Investment Corporation holds a 50 per cent share, Hongta Group 30 per cent, Yunnan Provincial Investment Corporation 10 per cent and the Huaneng Lancang River Hydropower Corporation 10 per cent (initially held by Yunnan Electricity Corporation). Before the establishment of Yunnan Dachaoshan Hydropower Corporation, the Water Resources and Hydropower Planning & Design General Institute and the Yunnan Provincial Development and Reform Commission jointly held a workshop to review the feasibility study for Dachaoshan in April 1990. The Ministry of Energy amended the feasibility study and in 1994 the National Development and Reform Commission approved it. In 1998, construction commenced. In 2000, the first generator began to produce electricity and all six generators had begun operation by 2003.

Emerging Huaneng Lancang River Hydropower Co., Ltd

The Huaneng Lancang River Hydropower Corporation emerged as a key actor and obtained all concessions on the Lancang mainstream dams except for Dachaoshan. It registered capital of RMB 4165.89 million (US$675.59 million), of which the CHG held 56 per cent, the Yunnan Provincial Investment Corporation 31.40 per cent, and the Hongta Group (a Yunnan-based company) 12.60 per cent. The CHG inherited the share from SPCC in 2002. In 2003, the CHG negotiated with the YPG and both parties signed a Memorandum on the Development of Yunnan Hydropower. The corporation's name changed from the Yunnan Lancang Hydropower Development Corporation to the Yunnan Huaneng Lancang River Hydropower Corporation, and then to the current Huaneng Lancang River Hydropower Corporation.

The Huaneng Lancang River Hydropower Corporation became a driving force of the Lancang hydropower development after it was incorporated into a shareholder company, which had its own legal entity and was an independent market player. Transfer of management from the Yunnan Electricity Bureau to the Huaneng Lancang River Hydropower Corporation established the boundary between administration and management as China's economic reform required. The YPG focused its efforts on administration and was responsible for policy formulation, implementation and monitoring, while the Huaneng Lancang River Hydropower Corporation was responsible for the interests of shareholders and seeking maximum returns on its investments. The renaming of the corporation without the prefix Yunnan semantically demonstrated that the CHG had become a primary shareholder with control over the corporation's operations.

A major legal framework was established to guide mega-projects such as hydropower project decision-making processes during the period of energy sector reform. Important laws were, for instance, the Water Law enacted in

1988, the Land Administration Laws enacted in 1986 and the Environment Protection Law enacted in 1989. At a project level, the NDRC is responsible for two major important decisions: the approval of the initial project proposal and the feasibility study. A developer has to work with other ministries before the project proposal and feasibility study are submitted to the NDRC for approval. The MoWR is responsible for technical approvals and the MoEP for environmental impact assessment (EIA) approval as examples. The decision-making process and documents required became more or less standardized after the establishment of the legal framework to guide mega-project decision-making. However, critics have pointed out that the standardized decision-making process is not transparent. EIAs are often handpicked by proponents and wider public participation is not allowed, only participation by a group of selected experts.[1]

The Nuozhadu approval process is an example. In 1999, the SPCC approved the draft feasibility study. The feasibility study was reviewed in 2002 and the Water Resources and Hydropower Planning & Design General Institute accepted the feasibility study. The MoWR approved the water and soil conservation plan in 2004. The State Environment Protection Agency approved the EIA in 2005. All of these documents were then submitted to the National Development and Reform Commission for its approval. The China International Engineering Consulting Corporation assisted the National Development and Reform Commission in the process of approving them. The author was informed that more than 70 permissions were required to develop and construct a hydropower project.

The emerging institutional arrangement has created favourable conditions in which to accelerate the development process. The Xiaowan, Jinghong, Gongguoqiao and Nuozhadu projects were prepared for in the late 1990s and early 2000s and then construction commenced in the 2000s. Xiaowan was the third mainstream project, with the second largest reservoir, and began operating at full capacity in 2010. Jinghong was the fourth mainstream project and its construction began in 2004. It reached full operational status in 2010. Gongguoqiao was the fifth mainstream dam; its construction commenced in 2007 and electricity generation started in 2011. Construction of Nuozhadu, which has the largest reservoir, began in 2006 and it began generating electricity in 2012. This rapid development can be attributed to the profit-seeking developer, the standardized decision-making process, local government cooperation, qualitative technical studies being prepared in advance and a supportive environment. This supportive environment also reflects the rhetoric of the major actors.

Hydropower justification and critique

There is a well-developed discourse in support of Lancang hydropower development that proclaims its benefits in terms of technical efficiency, economic advancement, clean energy and poverty reduction, modernization, civilization and enlightenment. This pro-hydropower rhetoric has obtained hegemony through speeches, the media, academic journals and government documents. The critiques are mainly presented in academic journals, which prudently raise concerns about the negative impacts on the environment and local livelihoods. This section will present Chinese debates regarding hydropower development.

Pro-hydropower rhetoric

The pro-hydropower rhetoric has emerged in statements from developers, consulting companies, high-ranking government officials and high-profile scholars connected to the government. It is worth noting that the deliberation of the scholars connected to government is very similar to that of the government itself because the former are often invited to argue in favour of the development initiatives proposed by the latter. Although each deliberation has its own emphasis, they share a common argument that uncritically promotes the benefits of hydropower while downplaying or ignoring many of the disadvantages.

Developers' rhetoric

Developers argue that electricity is crucial to support faster economic development. In 2004, Mr Kou, General Manager of the Huaneng Lancang River Hydropower Corporation, projected that demand for electricity in Yunnan would increase from 12,500 MW in 2010 to 23,480 MW in 2020, an increase of 87.84 per cent, while the demand for the West–East transmission will increase from 4,800 MW to 21,000 MW, a 375 per cent increase considering the average GDP growth of five provinces in southern China was 2.1 per cent greater than China's average (Kou, 2004). Xiao Peng, Chairman of the Huaneng Lancang River Hydropower Corporation Board, stresses that Yunnan has abundant hydropower resources that have barely been harnessed (Xiao, 2001). He has stated that Yunnan has a hydropower potential of 103,640 MW and a technical potential of 90,000 MW, accounting for 15.5 per cent and 17.9 per cent respectively of China's totals. Only 5 per cent is now in use, however.

Developers have also emphasized the social and environment contributions of hydropower. Wang Yongxiang, Vice President of the Board of the Huaneng Lancang River Hydropower Corporation, argues that energy development contributes to the modernization of a society (Wang, 2009) and that hydropower is a renewable and clean energy, a public good that can thus help build a 'jieyuexing shehui' or conservation-oriented society. Hydropower development is also promoted as a tool for poverty reduction, local economic development and the development of infrastructure. The Huaneng Lancang River Hydropower Corporation's efforts, he says, are also contributing to the success of the National Development Strategy, especially the China's Western Development Strategy.

Developers portray their arguments as 'common sense' because electricity is essential to sustaining economic development and can bring benefits to society and the environment. Developers' arguments supersede narrow technical understandings (Li, 2007), a process of delimiting the discussion in the technical field. Here we find that the developers also bring the discussions into the political and social fields. Politically, the argument aims to convince its audience that Lancang hydropower development will fulfil the National Development Strategy, which is deemed in the national interest and desirable for the people. Socially, hydropower is billed as renewable and clean energy that can contribute to building a conservation-oriented society. Needless to say, the developers' rhetoric dovetails well with the dominant discourse, at a time when the government is calling for either the implementation of the China's Western Development Strategy or the construction of a conservation-oriented society, and is thereby legitimatizes their actions and performance as supporters of state-led development.

Consulting companies' rhetoric

Consulting companies are in accord with developers and encapsulate their rhetoric by means of cost-effective rationale. Qing Changgeng, the former Vice-Chief Engineer of the HCKEC, has stressed the technical efficiency and competiveness of Yunnan hydropower. Yunnan's hydropower is, he explains, characterized by huge standing stock, wide distribution, rich and stable water discharge, a profound single purpose for hydropower development, favourable damming conditions due to low inundation areas, less civil work, lower construction costs and higher economic returns (Qing, 1997).

Hydropower development will also provide opportunities for local economic development and poverty reduction (Yang, 2001). Yang pointed out that 30 per cent of the investment would be spent locally on creating jobs, improving infrastructure and on the consumption of local materials. Local governments will benefit from additional revenue of RMB 6 billion from the Lancang's hydropower development. This revenue can then be deployed to support improvements in living standards, and on poverty reduction and environmental conservation. The electricity produced will supply local residents and can partially substitute for wood fuel, which will yield positive environmental consequences.

Consulting companies build their arguments on science and rational logic. The rhetoric aims to brand their technical excellence and appeal to developers, investors and governments. However, insufficient, if not missing, in their rhetoric are the negative impacts on the environment and local livelihoods. If the negative impacts were referred to, they could start to be addressed through technical means. Consulting companies simply assume that the economic benefits can unconditionally trickle down to the local communities. This is not true in the absence of proper institutional arrangements. For example, the local community can only use electricity instead of wood fuel if they can access and afford the electricity.

Government and scholars' rhetoric

Hydropower is purported to be a silver bullet for the economic development of Yunnan Province. According to the abstract of the 'Yunnan Provincial Tenth Five-Year Plan' (2001–5), the development of hydropower not only benefits the economic development of Yunnan Province but also supports that of other provinces. In 2009, the YPG published the Abstract on Yunnan Provincial Energy Development Planning (2009–15), in which hydropower remained the development priority. The YPG's governors and vice-governors constantly highlight the importance of hydropower development to Yunnan's economic development. He Zhiqiang, former YPG governor, stressed, 'the Lancang River Basin is an important economic zone' (Yunnan Electric Power, 1995). Niu Xiaoyao, former YPG vice-governor, pointed out that 'hydropower is a key sector', 'the West-East Electricity Transmission can advance the economy' and 'hydropower is a crucial means to upgrade the economic structure' (Niu, 1996).

Scholars have also been mobilized to justify the Lancang's hydropower development. One widely cited study incorporated the perspectives of the government, the developer, consulting companies and other knowledge to date and concluded that the abundant natural resources of the Lancang River Basin are a unique advantage for Yunnan to develop its economy (Yu *et al.*, 1988). The study stated that hydropower development could 'bianziyuan youshi wei jingji youshi' or 'turn resource advantages into economic advantages'. The study argued that the western part of Yunnan was poor; it had electricity shortages and needed environmental conservation. Therefore, development was urgently needed. It also proposed the formation of the Lancang River Development Corporation. The proposed corporation would be granted concessions for hydropower projects on the Lancang and could incorporate potential investors in addition to the financial support of central government. The assets of the Manwan project could be transferred to the proposed corporation in order to generate enough financial resources for further project development.

This rhetoric underscores the 'common sense' that economic development is essential to prosperity. Hydropower development is articulated as a fundamental sector to help harness natural resources, promote economic development and lift people out of poverty. This rhetoric emphasizes the positive impacts of hydropower development and understates the negative environmental and social impacts. Drawing attention to the negative impacts of hydropower does not mean taking an anti-development or anti-dam stance. Indeed, framing such observations in the light of trying to develop the best possible hydropower could help prevent future damage and identify better mitigation solutions, as discussed below.

Alternative rhetoric

Debates over the Lancang hydropower development are not as furious as that over the 'Three Gorges Dam', a cascade of dams on the Jinsha River (upper reaches of the Chang/Yangtze River) and Nu River (upper reaches of the Salween). A small group of environmental journalists, civil society activists and scholars have criticized the 'clean' and 'green' labelling of hydropower. They have also challenged the 'common sense' proposition that hydropower is essential to local economic development. They have raised concerns over the environment, social costs and justice of hydropower development.

Xiaohui Shen (2004), an environmental journalist, challenged the notion that hydropower is clean and green by comparing it to electricity produced by a coal plant. He referenced a Brazilian study that revealed that the soil inundated by a dam could also emit greenhouse gases. Shen argued that the greenhouse gases produced by a coal plant could be sequestered as technology advances, while those of a hydropower project could not be controlled. In addition to greenhouse gas emissions, hydropower can cause loss of biodiversity and induce earthquakes. Hydropower cannot, Shen asserted, be labelled simply as 'clean and green'. Hong Gao (2004), another journalist, pointed out that hydropower is not cheaper if the costs include public water use, biodiversity loss, environment degradation and loss of livelihoods.

Shen also argued that hydropower development barely benefits local people. It is not, he argued, sustainable development. A hydropower project may benefit the local economy in the short term, but because it relies heavily on the extraction of natural resources, it cannot be sustained in the long run. Many cases reveal that developers, investors and governments benefit from hydropower development while local people are negatively impacted. Xiaogang Yu (2005) provided evidence in support of the above argument. In 2001, Manwan made a profit of RMB 120 million, and paid tax of RMB 100 million to the central government, RMB 50 million to the YPG and another RMB 50 million to the county governments where it is located. But the people afected by the dam were still living in poverty.

Critics have also targeted the hydropower development process. A hydropower project may inevitably bring negative social and environmental impacts. A quality assessment can help to identify and predict both the positive and negative impacts, and solutions can be sought to mitigate the negative impacts. This author and other researchers have found, however, that the assessment often exaggerates the positive impacts and downplays the negative impacts in order to ensure that the project is approved. The current EIA system is called into question because consulting companies are contracted by developers to undertake the assessments and therefore have a conflict of interest in producing the EIA (Shen, 2004).

Although raising important points, the alternative rhetoric has failed to recognize the reality of China's energy features. Approximately 80 per cent of the country's energy sources are located in west China and 60 per cent of its energy consumption is in east China. In addition to coal and oil, hydropower is an important energy source that can be commercially harnessed while solar and wind energy alone cannot meet the increasing demand. Any hydropower project is inherently harmful to the environment and local livelihoods to some extent. However, proponents point out that the critics' rhetoric either focuses on or exaggerates the negative impacts and ignores the positive impacts. Some negative impacts are the result of mismanagement and are caused by factors other than the dam (Zhang, 2008). The alternative rhetoric can also be very biased and may not truly represent the interests of the affected people. Very often the two opposing rhetorics are built on a simplified and abstract 'world' and not on the complex and dynamic realities. The chapter now turns to understanding in greater detail the impacts of hydropower projects on livelihoods and the environment.

Impacts on livelihoods and the environment

A hydropower project will to some extent impact local livelihoods. The challenge is in how to mitigate negative impacts. As the first mainstream dam on the Lancang, the Manwan did not deal with resettlement well. Many lessons can be learned from the Manwan experience. Nuozhadu, the latest mainstream dam, has been labelled an 'environment-friendly dam'. In the analysis that follows, I draw on information about Manwan from previous studies, while information on Nuozhadu was collected from field investigations. The environmental impacts have been drawn from studies in academic journals.

The Manwan Dam's impacts on livelihoods

The Manwan Dam is 132 m tall, and has a reservoir capacity of 920 million m³. Its reservoir is 71 km long, and covers 23.9 km². It has an installed capacity of 1,500 MW. It was constructed in 1986 and started to generate electricity in 1993. As discussed above, the joint investment model employed to fund the construction of the dam was referred to as 'the Manwan Model'; it was also referred to as the most cost-effective project with the lowest investment per kilowatt-hour.

The filling of the dam's reservoir had a significant impact on the local people who relied mainly on farmland for their livelihoods. 7,260 rural people were resettled, of which 684 people moved out from the reservoir area, and 6,576 people moved up the reservoir, from the lowland to the upland. The reservoir inundated 241.92 ha of paddy fields, 173.04 ha of upland, 561 ha of forestland and 766 ha of bush (Yunnan University and Yunnan Provincial Manwan Electricity Generation Plant, 2000). Losses were due not only to the area of farmland inundated, but also to its productivity because lowland farmland (paddy and dry-land) is more productive than upland farmland.

The resettlement plan promised the resettled people that all losses would be compensated. In practice, this promise was hard to keep because equally productive farmland could not be found in the mountainous areas. The total compensated reclaimed farmland area (newly reclaimed land on slope) was 17.2 per cent less than the amount of farmland taken. Table 4.2 summarizes the various types of farmland losses suffered by resettled people as a result of the

Table 4.2 Land use before and after the construction of Manwan Dam

Farmland	Loss (ha)	Compensation (ha)	Change (ha)	Percentage of the change
Paddy	241.92	153.52	−88.40	−36.5
Slope rain-fed land	173.04	206.49	+33.45	+19.3
Total	414.97	360.01	−54.96	−17.2

Source: Fu and He, 2003.

creation of the dam's reservoir. To make the situation worse, the new irrigation systems, which were part of the resettlement plan, either lacked management or competed with existing irrigation systems for the same water (Zheng, 2012). One-third of the reclaimed paddy was later converted into rain-fed land (Chen and Li, 2003). The farmland offered as compensation was not as productive as the farmland lost to inundation. The yield from paddy yields, for example, was reduced by 3,000 to 4,500 kg per ha, while that of rain-fed agriculture declined by 1,500 to 3,000 kg per ha. This lower yield forced the villagers to claim 130 ha of additional farmland (Fu and He, 2003).

This loss was borne not only by the resettled people, but also by the villagers who hosted the resettled people. Most of the resettled people were reallocated land between 1,000 and 1,500 m above sea level. The local government required the host villagers to redistribute their farmland in order to secure the livelihoods of the resettled people. The land was reallocated equally between the hosts and the resettled people, so as to reduce the resettlement costs (Zheng, 2012).

Access to other resources was also reduced. In 1998, the Manwan management decided to form a contract with a third party to raise fish in the reservoir. Villagers along the reservoirs were not allowed to fish any more. The contractor hired guards to prevent the villagers from fishing. However, this initiative failed after three years (Zheng, 2012). Many of the affected people also had difficulties raising cattle and collecting wood for fuel because a lot of land had been developed into farmland. 'Electricity for fuel wood', a claim in support of hydropower, was shown to be incorrect, simply because the affected people could not afford the price of electricity (Zheng, 2012).

Intensified land use and fluctuation of the reservoir's water level increased natural disasters. One study by Fu *et al.* (2005) reveals that soil erosion increased in the resettlement areas. In some places, serious soil erosion affected between a third and a half of the total area; the number of landslides, the debris flow and flooding along the reservoir increased by 30 per cent after the dam's construction. Within the first three years (1993–6), the sedimentation depth in the reservoir had increased by 30 m, equivalent to five years' deposit as estimated in the original dam design.

As a result of all of these difficulties, incomes declined among the resettled people. In 1991, their average annual income was higher than the county average and by 1996 it was lower (Chen and Li, 2003). Over 2,030 people, nearly a third of those resettled, were affected by landslides, shortage of water supply, and increased damage to farmland. These consequences forced the YPG to resettle them for a second time in 2003 (Zhou and Fu, 2008). The dam also created a variety of tensions between the people who were resettled by the dam project and those who hosted the resettled people, as well affecting their relations with local government (Zheng, 2012). Although many issues were later resolved, the lessons remain for future hydropower development.

On reflection, four factors are responsible for the inadequate way in which the social impacts of the Manwan were addressed. First, the resettlement plan was rushed. Many issues were not properly assessed and discussed among the stakeholders. The affected people were not consulted with respect to the resettlement plan. Second, the budget was very limited. The YPG committed a lump sum package of RMB 17.6 million for the resettlement plan implementation, although the practical costs later tripled. With this limited budget, many issues could not be properly addressed. Third, local government was in a weak position to fully protect the interests of the resettled people because of the limited budget and political pressure. Fourth, the legal system was not well-developed to guide hydropower development when the Manwan Dam was being prepared for and implemented.

The Nuozhadu Dam's impacts on livelihoods

The Nuozhadu Dam is 216.5 m tall. It has an installed capacity 5850 MW. Its construction began in 2006 and the first generator started to produce electricity in 2012. The remaining eight generators started to supply electricity in 2014. It has a reservoir capacity of 23,703 million m³, which covers an area of 320 km². The reservoir is 215 km long. It is claimed to be a green or 'environmentally-friendly' project (Yuan, 2012).

The lessons learned from Manwan seem to have been applied in the Nuozhadu's resettlement plan. The plan takes the view that the resettled people can 'maintain the same living standards after resettlement implementation' (Xu and Li, 2005). The resettlement implementation seeks to achieve the goals of 'bandechu wendezhu nengzhifu huanjingdedaobaohu' or the smoothly moving out, stably staying at the resettled places, better livelihoods than before and for the environment to be protected. It was estimated that 43,602 people would be resettled, of which 95 per cent would be rural people. The reservoir would inundate 5816 ha of farmland (paddy and rain-fed), 2613.5 ha of orchards, 17,994 ha of forest and 621.5 ha of pasture (Xu and Li, 2005).

Options have been made available for the resettled people to choose from. For instance, Pu'er City created a pamphlet and distributed it to the resettled people to explain the resettlement policy.[2] The resettled people were provided with three options for residential resettlement: 'kaohou banqian' or moving up the reservoir into concentrated residential areas; 'yidi banqian' or moving out of the reservoir into concentrated residential areas; or moving by themselves. The first two options are eligible for resettlement with support programmes:

long-term compensation and land compensation. Those who choose long-term compensation will receive a monthly payment of RMB 178 (US$30) and an additional 0.02 ha of paddy field per person. The monthly payment will be adjusted every two years depending on inflation. Those who choose the land compensation option will be compensated with land equivalent to that owned before the resettlement. In Pu'er Prefecture, follow-up production support will provide an additional RMB 4,000 per capita subsidy and an RMB 50,000 loan, which will be interest free for five years.

Our study selected 116 sample households to find out about landholding, income structure, the environment and other concerns. The sample households were located in five villages. In Lincang Prefecture, 26 households were sampled in one moving-out village and 30 households from a moving-up village. In Pu'er City, 30 households were sampled in a moving-out village and another 30 households from two moving-up villages.

The survey revealed that the landholdings among the sample of resettled people had reduced significantly. The moving-up people had lost all of their paddy field, accounting for 0.065 ha per capita. Other land (rain-fed land, orchards and forest) had decreased from 0.587 ha to 0.377 ha, or a 37.8 per cent decline per capita. All of the sample households had chosen the long-term compensation scheme and were not eligible to have paddy field compensation. The paddy land of the moving-out people had reduced from 0.052 ha to 0.035 ha, a decline of 32.7 per cent per capita. Their other land had decreased from 1.39 ha to 0.68 ha, a decrease of 51.08 per cent per capita, and this is located in their original villages, far away from the current residential areas. Those who had chosen land compensation were still waiting for their other farmland to be allocated.

The survey revealed that the people had reduced their dependence on agricultural activities and had a more diverse income structure than before. Their agricultural income had significantly reduced and they relied more on the long-term compensation and seasonal labour (Table 4.3). It is interesting to note that the moving-out people were seldom engaged in small business while the opposite is true of the moving-up people. Existing social networks play an important role here: the moving-out people had lost their social networks when they moved into the new residential areas.

Table 4.3 The income structure of the sample households

Type	Resettlement	Agriculture (%)	Seasonal labour (%)	Compensation (%)	Business (%)
Moving-up	Before	82.0	16.6	0	3.4
	After	36.7	22.7	26.0	15.6
Moving-out	Before	92.0	6.2	0	1.8
	After	42.3	37.3	22.4	1.7

Source: field survey in July 2013.

The Nuozhadu hydropower project has so far been better able to manage the physical environment around resettlement and farmland areas compared with Manwan. Approximately 70 per cent of the resettled people thought that they had a better natural and built environment around them, 12 per cent of them thought that there was no change and 18 per cent of them thought it was worse. Those who thought their surrounding environment was worse are people from the moving-out group. They also complained about a lack of wood for fuel and not having enough space to air-dry their grain. The interviewed officials claimed that the process of the people moving out had had been smooth. According to a resettlement expert who had assessed the resettlement plan, the implementation of the resettlement plan had stood out as a best practice case in China. These facts seems to confirm that the project has achieved its objective of smooth resettlement. It is, however, still too early to say whether or not the current resettlement scheme can achieve the goal of people 'stably staying at resettled places' and enable the people to have 'better livelihoods than before'.

Several issues emerged during the survey. The moving-up people still hope to have paddy fields although they are not eligible. Food security may be an issue in the coming years. The transportation network does not meet their expectations because of the increasing number of people doing small business. The moving-out people cannot fully take up new opportunities, such as small businesses, because they lack a social network. In addition, they often compete for water from the same irrigation system for their paddy fields. The existing management system is still in the process of adjusting and accommodating the newcomers. Local labour wages have reduced slightly because more people are competing in the local labour market. Finally, there are tensions between the original villagers and the resettled people.

The resettlement practices of the Manwan and the Nuozhadu projects demonstrate that good policy and implementation can produce very different outcomes. Critics often cite tManwan as a negative example of hydropower development. Despite not being perfect, Nuozhadu proves that hydropower can do better in the social field.

Environmental impacts

Dams can alter the spatio-temporal characteristics of the water environment. The surface water temperature of the Manwan reservoir has increased by 4.8°C. A similar change has also been observed in the Dachaoshan reservoir (Yao *et al.*, 2006). Data collected from 1988 to 2002 show that there was no change in the water quality of the Manwan downstream while that of the Manwan upstream decreased slightly due to an increase in pollution discharge (Zhang *et al.*, 2005). Between 1987 and 2003, sedimentation concentration at Chiangsaen decreased up until 1997 and then started to increase, while the concentration at Jinghong consistently decreased (Fu *et al.*, 2006).

The impacts of dams on flow are very dynamic. Analysis of monthly run-off data (1956–2001) revealed that the downstream flow was significantly disturbed during the construction of the Manwan and Dachaoshan dams but it was restored to its annual and monthly behaviour patterns after construction. However, the daily and weekly patterns remain disturbed (He *et al.*, 2006). The water discharge variation within a year also became slightly less sharp after the construction of the Manwan and Dachaoshan dams (Huang and Liu, 2010). It was estimated that the average flow in the lowest month, either April or May, would increase from 434 to 1,750 m^3/s and the dry season flow from 771 to 1,652 m^3/s after the construction of the Xiaowan and Nuozhadu (Zheng, 2004).

The dams also have impacts on biodiversity. The Lancang River Basin is a biodiversity hotspot with 80 plants and 131 animals on the national protection list. Critical scholars argued that the dams' inundation, especially that of the Xiaowan and Nuozhadu dams, could destroy some environmental habitats (Dai, 1995). The Manwan Dam has had a greater impact on upstream habitats than on downstream ones in terms of effect intensity, continuity and scale. A single dam's impacts may be within environment tolerances, while cascades of dams can result in 'immeasurable impacts' (Cui and Zhai, 2008).

The fish species and subspecies in the Lancang River account for 41.8 per cent of the total species of Yunnan and 20 per cent of the total fresh water species of China (Dai, 1995). There was a decrease in indigenous species and an increase in exotic species during the 1960s, 1980s, and 1990s. The hydropower development on the mainstream and tributaries of the Lancang River is one of many causes of this (Liu *et al.*, 2008). Zhang (2001) reported that the Manwan Dam has significantly affected the number of species in the river.

Studies to date reveal that dams on the Lancang have various social and environmental impacts. These dam impacts are not only recognized within China as presented above, but also at the international level (WCD, 2000). Regarding the Lancang mainstream dams, it seems that Chinese experts agree that the environmental impact of a single dam may be tolerable, but that of cascade dams cannot be ignored and require complex mitigation measures (Cui and Zhai, 2008). Tullos *et al.* (2009) further argue there is a need to document the interconnected nature of biophysical, socio-economic and geopolitical effects. Although there has been progress on the resettlement plan and implementation for the Nuozhadu, the interviewed critics pointed out that time is needed to assess its outcomes. Despite the various impacts and uncertainties, the development of the Lancang's hydropower is still ongoing. It is worth examining its institutional arrangement and drivers.

Institutional analysis

Hydropower is a way to harness natural resources for economic return, which competes with existing ways that communities and ecosystems use natural resources. This competition takes place at a local level and involves the developer, local governments and communities. The decisions embedded in institutional provision, consisting of rules, interest groups and power relations, are made beyond the local level. Magee (2006) has developed the concept of the 'powershed' to capture the complexity and dynamics of hydropower development in Yunnan from a perspective of politics of scale, and to examine the interaction of policy, the energy industry and governments at many levels. This section sets out to unpack the driving forces of hydropower development on the Lancang mainstream in terms of institutional arrangement and to examine the various power relations.

Controls over hydropower development

The state owns water resources and controls the development of large hydropower projects. The ownership of water resources is defined in Article 3 of the Water Law. The State Council is responsible for the execution of this claim on behalf of the state. Article 7 of the Water Law also stresses that the state shall apply a system of water permits and acquisition to the value to water. The Ministry of Water Resources is the authority responsible for implementing the system. The authority is required to take the leading role in integrated basin development planning and sector planning for critical rivers and lakes such as the Lancang River. The law also states that the State Council should promulgate regulations to guide the construction of medium and large dams. The land administration law reinforces the state's control over hydropower development through land acquisition. In China, land is either owned by the state or collectively by a group of rural people. The land law states that no one can acquire collectively owned land except the state. Such purchases are subject to State Council approval if the amount of land taken is greater than 35 ha for preserved farmland, and 70 ha for cultivated land. These are not the only instruments that the state can use to achieve its development goals.

Instruments of state-led development

China is a developing state and takes the necessary policy measures to achieve its economic objectives (Sun and Xu, 2009). The five-year plan is the most important instrument to achieve economic development. This planning takes place at the central, provincial and county levels. Mega hydropower projects must first be integrated into a five-year plan before they can commence. Prior to integration into a five-year plan, a large-scale hydropower project must be included in an integrated basin development plan and a hydropower sector plan in order to secure its economic and technical viabilities. The Manwan hydropower project was first identified in the Middle Lancang Plan and then included in the five-year plan. The former plan is a scientific process that engages the MoWR, the YPG and consulting companies. The latter is a political process that engages the NDRC, the MoWR, the YPG, the developers and consulting companies or scholars. At the project level many approvals are needed, for instance, proposal approval, feasibility study approval, EIA approval and resettlement plan approval. The author was informed that permission must be obtained from more than 70 institutions before a large-scale hydropower project can commence.

Inter-provincial institutions have emerged to carry out regional development initiatives beyond the provinces. A five-year plan is a process through which provinces can compete for favourable conditions from the central government with respect to the development of mega-projects such as hydropower projects. The competition can force provinces to roll their development aspirations forward. Such territory-based development plans can make economic development fragmented due to political boundaries and weakened economic integration beyond the provincial scale. There was a need for inter-provincial institutions to carry out inter-provincial development such as the China's Western Development Strategy. In 1999, the central government initiated a 'zhuada fangxiao' or 'developing the big companies and loosen the small companies' programme, which aimed to support the development of large corporations, especially SOEs, and liberate the market for small firms. The State Power Corporation of China (SPCC) was formed and then split into 11 corporations, including five energy-generating corporations, two electricity transmission and delivery corporations, two consulting corporations and two corporations. The 'zhengqi fengkai' or 'separation of administration from management' policy demarcates the boundary between government and the SOEs. The former focus more on policy formulation and implementation while the latter are granted market muscle to carry out the mission of economic development.

Approaches to hydropower market

The Chine Huaneng Group (CHG) quickly strengthened its position in the development of the Lancang hydropower after splitting from the SPCC. It immediately incorporated former investors as shareholders of a subsidiary, who had a dominant share of 56 per cent. The subsidiary was given a new name, the Huaneneg Lancang River Hydropower Corporation, without the prefix of Yunnan, to symbolize that it was an independent corporation responsible to its shareholders. Backed by the government policy which encouraged large-scale developments, CHG obtained a development concession for the Lancang and then negotiated with the YPG on the detailed development arrangement.

The Huaneng Lancang River Hydropower Corporation secured access to the market and a return on its investments. China Southern Power Grid Corporation, that split from SPCC, is a dominant buyer and supplier of electricity within Yunnan, Guizhou, Guangxi, Hainan and Guangdong. China's West Development Strategy helps these two corporations to work together to capture emerging market opportunities. The China Southern Power Grid Corporation transmits the electricity that the Huaneng Lancang River Hydropower Corporation generates to Guangdong province. The Huaneng Lancang River Hydropower Corporation was also permitted by the NDRC to sell electricity on-grid at a price that maintains an 8 per cent operation profit margin. The price is sufficient to secure a return on its investment.

Stable cash flow and huge returns on its investment have put the Huaneng Lancang River Hydropower Corporation in a better position to attract financiers.

The Huaneng Lancang River Hydropower Corporation can easily access various state-owned banks. For example, the Xiaowan project was financed by the China Development Bank, the Construction Bank of China, and the Industry & Commercial Bank of China (Dore *et al.*, 2007). It also issued a RMB 2 billion enterprise bond in 2010. The author was informed that there is no difficulty for Huaneng Lancang River Hydropower Corporation in accessing financial sources as long as its debt to asset ratio is less that 80 per cent, which is a policy for SOEs. Despite the favourable market environment, the Huaneng Lancang River Hydropower Corporation still needs the support of important partners, especially local governments, in order to smoothly implement hydropower projects.

Hydropower's partners

The development of a mega hydropower project is a formidable, if not impossible, task without the participation of consulting companies and local governments. Consulting companies provide their technical services either to the developers or to the administration. CIECC, WRHPDGI and HCKEC, like other large consulting companies in China, are also SOEs. They employ 200,000 registered engineers and have both technical capacity and strong connections with government.[3] These companies, in fact, have become part of the driving force pushing for mega-projects that help them earn a financial return, as well as supporting and enhancing their reputation.

Local governments are crucial partners for hydropower developers. Legal provision and economic benefits are push and pull factors for local governments. Article 29 of the Water Law states that hydropower developers must prepare a resettlement plan that is subject to approval by the authority. The implementation should be included in the project budget. Local governments are responsible for resettlement plan implementation. In 2006, the State Council issued a provision about large–medium dam resettlement management, which requires provincial governments to be responsible for the administration and monitoring of resettlement implementation and a county to be integrated into its implementation.

A hydropower project on the Lancang mainstream can generate tremendous financial profit that can be shared among governments. In 2001, the Manwan Dam brought RMB 50 million worth of revenue to the YPD and RMB 50 million to county governments (Yu, 2005). Hydropower is itself an important industry and can also contribute to the development of other industries. In Pu'er Prefecture Municipal City (consisting of nine counties and one district), for instance, hydroelectricity generation grew sixty-three-fold from 1986 to 2011, mainly from the Lancang mainstream projects. In 2011, hydroelectricity accounted for 15 per cent of industrial output and provided power for three other major industries: mining, manufacture and timber processing (Pu'er City Statistics Bureau, 2012). The benefits, however, come with environmental and social costs that need to be addressed.

Hydropower development safeguard policies

Much hope is placed on safeguard policies that can protect local livelihoods and the environment. The most important instruments of safeguard policies include resettlement plans, EIA and water and soil conservation plans. In the provisions of the Land Administration Law, land can be transferred to a developer via the government instead of the market if projects are subsidized by the state. For large–medium dams, the compensation rate for land is equal to 16 times the average output value of the past three years. The YPG has developed standard compensation rates for each type of land used at the county level. These provisions provide a standard reference to avoid conflict rates among different projects, and they leave no room for landowners to negotiate an increase in the compensation rate. The management of the Huaneng Lancang River Hydropower Corporation informed the author that they went to the highest limits of policy and legal provision to implement resettlement plans in favour of affected people. They cannot, however, break regulations and government policies.

The Environmental Protection Law and the EIA Law have made a progressive contribution to environmental protection in China (Zhu and Lam, 2009). The Huaneng Lancang River Hydropower Corporation also argues that the Nuozhadu project follows legal environmental provisions and policies. Critics have raised concerns that the EIAs should not be treated as a 'rubber stamp' (Dore *et al.*, 2007). During the field interviews, critics of Yunnan's hydropower development argued that there is insufficient biodiversity knowledge of the Lancang River. Licensed companies often lack personnel with specific knowledge on biodiversity. This situation calls into question whether they can produce a scientifically sound EIA assessment.

Compared to the Manwan Dam, the Nuozhadu project has made several improvements to protect the physical environment. It has adopted a stratified-intake water technology to maintain a normal temperature downstream; established fish breeding stations to maintain fish communities; and it has a botanic garden to preserve endangered plant species. The legal provisions also require the Huaneng Lancang River Hydropower Corporation to follow dam operation rules, for instance maintaining minimum flow. Interviewed experts pointed out that it is still early to assess the outcomes of these measures.

Power relations and hydropower

There are very dynamic relations between the developers, financiers, consulting companies, electricity distributors, local communities, civil society, and the government and its agencies at various levels (Dore *et al.*, 2007; McDonald *et al.*, 2009). Power relations among these actors have also evolved as China is transforming from a planned economy to a market economy.

With respect to the hydropower development on the Lancang mainstream, the NDRC is at the centre of decision-making and other ministries, especially the MoWR, the MoEP and the MoLR, play important roles in the process. There are three distinct groups that are trying to influence decision-making with regard to hydropower development. The Huaneng Lancang River Hydropower Corporation, hydropower consulting companies, the YPG and pro-hydropower scholars may be considered as associated proponents advancing the hydropower agenda. Local communities, civil society and critical scholars are deeply concerned about either livelihoods or environmental impacts, or both. Prefecture, county and township governments have to mediate between social stability and economic prosperity.

The associated proponents generate particular and powerful narratives that highlight the benefits of hydropower and its contribution to national and local economic development. They also encapsulate the negative impacts into a technical domain or technical rendering so that they can be resolved through engineering solutions. The legal framework also establishes the necessary conditions for the associated proponents to settle disputes over resources use and smooth project implementation. The public have limited access to crucial information regarding the impacts and therefore it is difficult for them to participate in meaningful dialogue. The associated proponents are well networked and can avail themselves of a mix of administrative, market, financial and technical power to maintain dominance.

Local communities, civil society and critical scholars are important forces that counter the associated proponents although they are not well networked with each other. Local communities can either appeal to local governments to represent their interests or sometimes take collective action, such as staging demonstrations to pressure governments and win the support of society. Local community actions can be regarded as 'buwending yinsu' or unrest elements. For instance, increasing demonstrations by local communities have received wide societal attention. Civil society, journalists and critical scholars can bring community concerns and technical evidence about negative impacts to the attention of a wide audience, which can generate social pressure on the associated proponents. Suspension of the dam on the Nu Jiang or Upper Salween River is an example.

Local governments are mediators who can influence on both groups. They can leverage the complaints of local communities and exert influence on the associated proponents to advance local interests. They can also avail themselves of the proponents' narratives to calm down uneasy local communities. They play a role of boundary organization, which coordinates the interests of developers and local communities as well as of its own so that a hydropower project can be implemented smoothly.

Conclusions and recommendations

China is a developing state following a state-led development model. It implements this partly through its administration and the management of state-owned companies. The state owns water resources and controls the use of other natural resources. The Lancang's hydropower development illustrates that the sector plan for hydropower, the national five-year development plan and project level approvals are the main instruments of administration. State-owned companies such as the Huaneng Lancang River Hydropower Corporation and the China Southern Power Grid Corporation incorporate the investment arms of central and provincial government that then become inter-provincial institutions that carry out hydropower development and exploit market opportunities because of the uneven distribution of energy sources and demands. The hydropower proponents are able to maintain the dominance of the pro-hydropower development rhetoric, availing themselves of central government development policies, technical knowledge and national economic benefits. Financially, sustainable returns of mainstream hydropower projects on the Lancang River make it possible to persuade lower level government agencies, consulting companies and pro-hydropower scholars to support hydropower development. These institutional arrangements are promulgated in laws and regulations as well as in polices. It is this combination of political, economic and market forces and the dominant pro-hydropower rhetoric that configures Yunnan as a 'powershed' and drives the hydropower development of the Lancang. The local communities have become the objects rather than the subjects of hydropower development and are excluded from the main decision-making process (Zheng, 2012).

Safeguard polices are tools that can protect livelihoods and the environment, constrain debates and negotiations, or legitimate hydropower development. Consulting companies can easily obtain licences for resettlement planning and EIA assessment, even though they may not have enough knowledge – especially with regard to biodiversity – to make a reasonable assessment. Although scientists warn that the cumulative impacts of hydropower cannot be ignored, there are still not enough studies or information available to assess the cumulative impacts of cascades of dams. Resettlement policies and implementation have been improved in the period between the Manwan and the Nuozhadu. More time, however, is still needed to evaluate the outcomes.

Dore and his colleagues (2007) suggest that China should revisit its energy policy and strengthen the approval and assessment process. It will be a long-term task for China to change the current institutional arrangement from favouring hydropower development to balancing hydropower with social and environmental costs. To achieve this change, there is a need for China to implement several actions. First, the cumulative social, political, economic and environmental impacts of cascade dams on the environment and livelihoods must immediately be assessed. This will develop the necessary knowledge to further assess policy and implementation procedures. Second, there is a need to make information available to the public. A huge amount of information has been collected and presented in planning, assessment reports and study reports. We were able to access some of this information to inform this chapter, but the majority of it is still unavailable to the public. Third, it is necessary to create an open, mutually respectful, evidence-based debate or deliberative environment. Civil society and affected communities should be allowed to participate in the decision-making process. Fourth, it is essential to establish a well-functioning international cooperation/institution to manage the international rivers and benefit both the upstream and downstream.

Acknowledgements

The chapter is the product of research implemented by the CGIAR Challenge Program on Water and Food (CPWF), with funding from Australian Aid. Nathanial Matthews, Kim Geheb of the CPWF, and Zha Daojiong of Peking University provided invaluable comments. The York Center for Asian Research (YCAR) of York University hosts the author as a visiting scholar and provided its incredible facilities to support the writing of this chapter. Peter Vandergeest shared his insights on political ecology. Editors helped a great deal to improve the writing. The author is grateful for their assistance.

Notes

1 Magee, 2014, personal communication.
2 Pu'er City Resettlement Bureau (2011), the compensation and resettlement policy.
3 Zha, D. J. (2013), personal communication.

References

Chen, L. H., and Li, Q. (2003) 'Obstacles in production restoration and related reasons in the area around the Manwan dam', *Resources and Environment in Yantze Basin*, 12(6): 541–6 (in Mandarin).

China Statistics Bureau (1996) *China Statistical Year Book 1995*, Beijing: China Statistics Press (in Mandarin).

China Statistics Bureau (2012) *China Statistical Year Book 2012*, Beijing: China Statistics Press (in Mandarin).

Cui, B. S., and Zhai, H. J. (2008) 'Quality evaluation of habitats disturbed by the Manwan hydropower dam', *Acta Scientiae Circumstantiae*, 28(2): 227–34 (in Mandarin).

Dai, L. (1995) 'Integrated development of the Lancang River and biodiversity conservation', *Yunnan Environment Science*, 14(4): 14–19 (in Mandarin).

Ding, L. (2005) 'Yunnan hydropower timeline 1910–2005', *Yunnan Electricity Power*, 11: 16–17 (in Mandarin).

Dore, J. X., Yu, G., and Li, K. Y. (2007) 'China's energy reforms and energy expansion in Yunnan', in L. Lebel, R. Daniel, and Y. S. Koma (eds), *Democratizing Water Governance in the Mekong Region*, Chiang Mai: Mekong Press, 55–92.

Fu, B. H., and He, Y. B. (2003) 'The effect on affected people's income and reservoir area ecology caused by farmland changes of Manwan Hydropower Station', *Territory and Natural Resources Study*, 4: 45–6 (in Mandarin).

Fu, B. H., Chen, L. H., and Zhu T. (2005) 'An analysis of ecological environment change in the storage areas of Manwan Hydropower Station and its managing countermeasures', *Territory and Natural Resources Study*, 1: 54–5 (in Mandarin).

Fu, K. D., He, D. M., and Li, S. J. (2006) 'The downstream sedimentation response to the development of the Lancang mainstream hydropower', *Chinese Science Bulletin*, 51: 100–5 (in Mandarin).

Gao, H. (2004) 'Challenges of sustainable development to hydropower industry', *Technology Estate*, 4: 31–2 (in Mandarin).

Gao, J. (1988) 'A workshop on integrated economic development strategies of the Lancang Basin held in Beijing', *Yunnan Water Power*, 4: 66–7.

He, D. M., Fen, Y., Gan, X., and You, W. H. (2006) 'Cross-border hydrologic effects of cascade mainstream dams in the Lancang River', *Chinese Science Bulletin*, 51: 14–20.

Huang, Y., and Liu, X. Y. (2010) 'Impact of hydropower development on annual runoff and sediment transport distribution within year', *Advances in Water Science*, 21(3): 385–91 (in Mandarin).

Kou, W. (2004) 'Accomplishing the mission and accelerating development to push Lancang hydropower construction into a new stage', *Water Power*, 30(10): 1–4 (in Mandarin).

Li, T. M. (2007) 'Practices of assemblage and community forest management', *Economy and Society*, 36(2): 263–93 (in Mandarin).

Lin, Z. G. (1993) 'A hydropower development timeline of the Lancang River', *Yunnan Water Power*, 2: 93–5 (in Mandarin).

Liu, Y. H., Huang, Sh. L., Chen, J. H., and Yan, X. G. (2008) 'Strategy study on regional conservation and management of fish resources in the Lancang-Mekong River Basin', *Journal of China Agricultural University*, 13(5): 55–62.

Liu, Z. W., and Chen, Q. P. (1992) *Manwan Dynamics*, Kumming: Yunnan Nationality Press (in Mandarin).

Ma, H. Q. (2003) 'Current, future and suggestions for Yunnan hydropower development', *Yunnan Power Industry*, 11: 7–10 (in Mandarin).

Ma, H. Q. (2004) 'Exploration and analysis of hydropower development prospects of the Lancang River Basin', *Yunnan Water Power*, 20(5): 1–4 (in Mandarin).

Magee, D. (2006) 'Powershed politics: Yunnan hydropower under Greater Western Development', *China Quarterly*, 185: 23–41.

McDonald, K., Bosshard, P., and Brewer, N. (2009) 'Exporting dams: China's hydropower industry goes global', *Journal of Environmental Management*, 90: 294–302.

Niu, S. Y. (1996) 'Completion of Daochaoshan and advancing Yunnan economy to new stage', *Yunnan Water Power*, 2: 1–4 (in Mandarin).

Pu'er City Statistics Bureau (2012), *Pu'er Statistics Year Book 2011*, Pu'er: Pu'er Prefecture Municipal City Statistical (in Mandarin).

Qing, C. G. (1997) 'Harnessing hydropower potential and advancing the Yunnan economy with the assistance of West–East Electricity Transmission', *Water Power*, 1: 9–11 (in Mandarin).

Shen, X. H. (2004) 'Ten questions about the Nu River', *Green China*, 2: 8–16 (in Mandarin).

Sun, P. D., and Xu, J. N. (2009) 'From miracle to crisis: Theory of developmental state and beyond', *Social Sciences in Guangdong*, 2: 173–8 (in Mandarin).

Tullos, D., Tilt, B., and Liermann, C. R. (2009) 'Introduction to the special issue: Understanding and linking the biophysical, socioeconomic and geopolitical effects of dams', *Journal of Environmental Management*, 90, supp. 3: S203–S207.

Wang Y. X. (2009) 'Knowledge and the practice of sustainable development of river basin hydropower resources', *Water Power*, 35(9): 5–9 (in Mandarin).

WCD (World Commission on Dams) (2000) *Dams and Development*, London: Earthscan.

Xiao, P. (2001) 'Seizing opportunity and advancing Lancang hydropower development', *Yunnan Electricity Power*, 2: 27 (in Mandarin).

Xu, Y., and Li, H. Y. (2005) 'Rural resettlement of Nuozhadu hydropower station', *Water Power*, 31(5): 20–2 (in Mandarin).

Yang, R. R. (2001) 'The leading role of Yunnan hydropower in national "West to East Electricity Transmission Strategy"', *Yunnan Water Power,* 17(1): 1–5 (in Mandarin).

Yao, W. K., Cui, B. S., Wei, S. K., and Liu, J. (2006) 'Spatio-temporal characteristics of Lancang Jiang River water temperatures along the representative reaches disturbed by hydroelectric power projects', *Acta Scientiae Circumstantiae,* 26(6): 1031–7 (in Mandarin).

Yu, G. Y., Gao, Z. G., Lin, H., Luo, X. B., and Zhu, K. (1988) 'A comprehensive study tour report on the development of the hydropower economy of the Lancang River Basin', *Technological Economy,* 5: 1–4 (in Mandarin).

Yu, X. G. (2005) 'Participatory social impact assessment on Manwan hydropower project', in Y. S. Zheng (ed.), *Scientific Valley Development,* Beijing: Huaxia Press (in Mandarin).

Yuan, X. H. (2012) 'Planning and practice on green hydropower of Nuozhadu', *Water Power,* 38(9): 5–8 (in Mandarin).

Yunnan Electric Power (1995) 'Governor He talks on electricity: Focus on large-scale grid coverage for electric development', *Yunnan Electric Power,* 24(17): 12 (in Mandarin).

Yunnan University and Yunnan Provincial Manwan Electricity Generation Plant (2000) *Environment and Biologic Resources of Yunnan Provincial Lancang River Manwan Hydropower Station Catchment,* Kumming: Yunnan Scientific Press (in Mandarin).

Zhang, B. T. (2008) 'Speeding up of hydropower development is current urgent task in practicing science-oriented outlook', *Sichuang Water Power,* 27(S1): 134–43 (in Mandarin).

Zhang, R. (2001) 'Review on evaluations of the ecological environment of the Manwan Hydroelectric Station on the Lancang River', *Hydroelectric Power Station Design,* 17(4): 27–32 (in Mandarin).

Zhang, Y. X., Liu, J. Q., and Wang, L. Q. (2005) 'Changes in water quality in the downstream of Lancang Jiang River after the construction of Manwan Hydropower Station', *Resources and Environment in the Yangtze Basin,* 14(4): 501–6 (in Mandarin).

Zheng, H. (2012) *Nature-Culture-Power: An Anthropological Investigation on the Manwan Debate,* Beijing: Copyright Press (in Mandarin).

Zheng, J. T. (2004) 'The roles of Lancang mainstream hydropower projects in the economic development of Yunnan', *Yunnan Water Power,* 20(5): 18–20 (in Mandarin).

Zhou, J. H., and Fu, B. H. (2008) 'Cultivated land use in the migrant areas of large-sized hydropower stations in Yunnan Province: A case study of the Manwan Dam', *Tropical Geography,* 28(6): 551–4 (in Mandarin).

*Zhu, T., and Lam, K. C. (2009) *Environment Impact Assessment in China,* Tianjing: Nankai University and Chinese University of Hong Kong, retrieved March 2014 from http://cseac.grm.cuhk.edu.hk/publications/EIA_IN_CHINA.pdf.

5 From Pak Mun to Xayaburi

The backwater and spillover of Thailand's hydropower politics

Jakkrit Sangkhamanee

Introduction

In January 1997, the street in front of Government House in Bangkok was occupied by a large crowd of villagers from along the Pak Mun River in northeastern Thailand. Mobilized under a nationwide movement called the Assembly of the Poor, the villagers' immediate goal was to submit a petition to then Prime Minister Banharn Silpa-archa demanding that his government set up an independent committee to oversee the problems arising as a result of the construction of the Pak Mun Dam, the state-of-the-art hydropower project completed in 1994 and operated since then by the Electricity Generating Authority of Thailand (EGAT). The villagers' concerns included issues of community relocation, compensation for the loss of their fisheries, houses and agricultural land, and EGAT's unfulfilled promise to develop extra income-generating activities for affected communities and to address the dam's impacts on the local ecosystem of the Mun River. The protest lasted for 99 days and the villagers established a temporary village on the street and pavements that cut through Bangkok's administrative area.

It was not until the government finally promised to take action regarding their concerns that the villagers agreed to end their street occupation. This event, however, was just part of a long struggle about the Pak Mun that has lasted 24 years. With 13 prime ministers and 16 government administrations involved in interminable and incoherent decision-making, even today, the conflict over the management of the Pak Mun Dam is far from over. In the early days of July 2013, the Pak Mun people, under a new alliance called the People's Movement for a Just Society, again assembled in front of Government House. This time they demanded that the Prime Minister, Yingluck Shinawatra, quickly set up a committee to try to resolve the disputes regarding the opening of the dam's sluice gates and the unresolved compensation problems that have long been endured by the local communities.

Since its initial feasibility study during the 1970s and its formal inception in 1989, the Pak Mun Dam 'issue' has become both *cause célèbre* and governmental *bête noire*. The Pak Mun controversy has been complicated by an obscure policy path, which has forced the Pak Mun villagers to take to the streets as the only

Figure 5.1 Thai NGOs campaigning against the construction of the Xayaburi Dam in Laos
Source: photo courtesy of International Rivers, Thailand.

effective means to directly negotiate with the decision-makers and oblige them to consider their problems.

The streets of Bangkok are not the only battleground in this contest. The Mekong River also provides a space for such contestation (Molle *et al.*, 2009). The view from Laos's capital city, Vientiane, looks directly across the Mekong into Thailand's Nong Khai Province. It was from the Nong Khai shore that, in November 2012, a flotilla of small fishing boats carrying more than 300 protesters set off to ply the Thai waters in front of Vientiane. The group, who called themselves the 'People's Network of Mekong Riparian Provinces' held aloft banners that read 'Stop Xayaburi Dam', 'Save the Mekong' and 'Say No to Xayaburi'. Their intention was to catch the eye not only of the government officials ensconced in discussions regarding the Mekong River Commission's (MRC) 'Procedures for Notification, Prior Consultation and Agreement' (PNPCA) at the Commission's secretariat in Vientiane, but also of those at the Asia-Europe Meeting (ASEM) Heads of State meeting. At this juncture, getting attention from political leaders, not just in Mekong Region but also from visiting countries, was an escalation deemed necessary. This was the first time that the MRC's PNPCA had been invoked, and it was this that the protesters sought to influence. The campaign attracted people from the eight Thai provinces that border the Mekong, as well as from areas where the impacts of the Xayaburi Dam are expected to be felt.

Despite the protest from the other side of the border, the Xayaburi Dam's groundbreaking ceremony was held after the ASEM concluded. The event was presided over by Laos's Deputy Prime Minister, Somsavat Lengsavad, who read out a statement saying: 'we had the opportunity to listen to the views and opinions of different countries along the river. We have come to an agreement and chose today to be the first day to begin the project' (quoted in Chenaphun, 2012). Notwithstanding that this mega-project is being built in Laos, Thai-based public companies (Ch. Karnchang, PTT and EGGO) are among the major shareholders of the dam's operator, the Xayaburi Power Company. Under the plan, approximately 95 per cent of the hydropower produced by the Xayaburi Dam will be sold to the EGAT to meet rising electricity demands in Thailand.

The two events described above – the street and river protests – focused on what were perceived to be shortcomings in dam decision-making processes, low levels of public participation in hydropower management, and the fears that the same litany of dam-related environmental and social problems would simply repeat themselves. I argue that the Pak Mun case did very little to change the way in which dam decisions were made in Thailand, and hence, I refer to it as a 'backwater' in domestic resource management, development discourses, knowledge production as well as stakeholder relations in Thai hydropower decision-making processes. The case, however, did herald a change in the strategies deployed to initiate, locate, plan, manage and operate dams outside the country.

In this chapter, I propose that the shift in Thailand's hydropower dam construction to its neighbouring countries has to a great extent been influenced by social and political tensions emerging as a result of incompetent and uncoordinated decision-making in the country's hydropower sector. Because of the inertia in this backwater, no effort has been made to reform the way in which decisions are made, forcing the sector to seek electricity supplies outside the country, in effect 'spilling over' its own internal shortcomings. Employing the two hydrological analogies of 'backwater' and 'spillover' to conceptualize and problematize hydropower development in Thailand, I propose that little has changed since the construction of the Pak Mun Dam with regard to improving the governance and regulatory practices to reduce the negative impacts of dam construction. A closer examination of the recent development of the Xayaburi Dam reveals that the transnational power play among hydropower stakeholders is becoming more complicated, however. The new hydropower landscape involves multi-sectoral actors as well as transborder issues that create new challenges with regard to understanding and improving the ways in which Thailand's hydropower is developed.

The political ecologies of backwater and spillover

Political ecology is a practical approach to analyse hydropower politics in Thailand and the Mekong Basin. Hitherto, studies on the politics of the Mekong's hydropower development have tended to explore this subject matter from a regional perspective (cf. Hirsch, 1999; Greacen and Palettu, 2007; Middleton *et al.*, 2009). The work by Middleton *et al.* (2009), for example, while successfully portraying the complex relationships between hydropower stakeholders at the regional level, touches very little on the domestic and local actors who play crucial roles in the decision-making processes. The difficulty of researching local actors is possibly one of the factors that hinders the cross-scale analysis of decision-making processes in Thailand and the neighbouring countries. An exceptional case is Foran and Kanokwan's (2009) Pak Mun Dam study, in which detailed information – regarding the development of the project, conflicts, the multi-layered stakeholders involved in the long battle of the negotiations, and the variety of campaigning strategies deployed by each of the actors to gain access and make convincing arguments in the decision-making process – is laid down systematically and chronologically. Foran and Kanokwan (2009, 74–5) argue that 'a dam planned and implemented with low transparency and accountability helped to trigger an unfolding, emergent series of disputes'. The case of the Pak Mun Dam 'offers sobering lessons about politics of knowledge' in which 'knowledge production did not always contribute in a "rational" way to inform negotiation'. It is important to look closely at the way in which knowledge and decisions in hydropower development are processed, articulated and shaped by centralized authorities within the circle of government and development enterprises. Analysing these forces helps to shed light on the drivers and enablers of the decision-making processes.

Some scholars argue that, if the decision-making processes in hydropower development are to be improved, it is essential for civil society to 'democratize an authoritarian state' (Foran and Kanokwan, 2009, 74–5) through, for example, street protests or by 'counter-framing' dominant knowledge (cf. Lebel *et al.*, 2007; Jakkrit, 2012). Building upon Foran and Kanokwan's (2009) analysis, this chapter examines the various strategies that different stakeholders have used to counter-frame the decision-making processes and dominant knowledge in the cases of the Pak Mun and Xayaburi dams.

The development of any large-scale hydropower project in Thailand now requires an environmental impact assessment (EIA) prior to construction. When the Pak Mun Dam was constructed in 1990, an EIA was not legally required. It was only after the promulgation of the Enhancement and Conservation of National Environmental Quality Act (ECNEQA) in 1992 that the environmental quality monitoring and evaluation system came into effect under the supervision of an independent committee (Thavivongse, 1998). The Act also allowed for more participatory roles for the local administrative authority in the control, prevention and resolution of environmental impacts. To a certain extent, the ECNEQA emerged as a result of the environmental movement, supported by the civil society organizations that had participated in the United Nations Conference on the Human Environment in Stockholm in 1972 (CUSRI and ERI, 2001). The event led to a movement in Thailand that called for an environmental protection law, which in 1975 became the first ECNEQA. This law was amended several times and re-enacted as the current ECNEQA in 1992.

In addition to environmental regulation, energy management in Thailand falls under multiple state agencies. The main organizations are the National Energy Policy Council (NEPC), the governmental organization that oversees the direction and investment strategies of the country's energy development, and an independent organization called the Energy Regulatory Commission (ERC) that was established by the 2007 Energy Industry Act. These organizations are the main administrative authorities that decide on the direction of future energy development in (and for) the country. The NEPC was established as part of the Sixth National Economic and Social Development Plan, at a time when Thailand was experiencing several energy crises. A prominent lesson derived from the Pak Mun case was that overall energy-related administration lacked unity among its authorities, both within the bureaucracy and among independent organizations (CUSRI and ERI, 2001). This was partly because these organizations were under the supervision of different offices and operated according to different chains of command. Moreover, the government still lacked a permanent mechanism for energy planning and had clarified neither its own role nor that of the private sector. Furthermore, it had not coordinated assistance between state organizations to ensure administrative continuity and efficiency (Greacen and Greacen, 2004).

Besides the above authorities, the National Economic and Social Development Board (NESDB) and EGAT have also played a crucial role in the hydropower decision-making processes. Their roles are to lay out fundamental hydropower policy and infrastructure, but they also dominate the construction of authoritative knowledge and social discourses on how water resources should be utilized and how electricity should be produced to serve the country's energy security and economic demands. Their instrumental rhetoric often highlights the ineluctable construction of hydropower dams, fast-growing national energy demands and anticipated future electricity supply shortfalls. This rhetoric is now widespread in the media and public knowledge, despite the fact that there are studies that challenge the view that there is an inexorable increase in energy demands, problematizing the means by which EGAT calculates its projections as well as its outdated energy efficiency policies and strategies (Greacen and Greacen, 2004; Greacen and Palettu, 2007). It has been argued, indeed, that such state rhetoric on energy security 'has often been used as an excuse for governing elites to pursue centralized industrialization and grandiose energy projects at the expense of marginalized population' (Simpson, 2007, 539).

The recent change in hydropower development in Thailand and the move towards its neighbours is not, however, the result of decisions made by the state authorities per se. As Bryant (1992, 18) reminds us, 'state policies are not developed in a political and economic vacuum. Rather, they result from the struggle between competing actors seeking to influence policy formulation' (cf. Bryant, 1999). In this sense, if we are to understand the power plays within the contesting space of hydropower development, it is important to investigate the involvement of multiple actors from different levels – ranging from local and national to regional and perhaps global – to see how these actors act to influence the process of decision-making and the outcome of given policies. The challenge here, as Bryant (1992, 18) notes, is to identify the 'different and often conflicting pressures on policy-makers' as well as to explore 'how previous policy choices contributed to the environmental change, and how such change in turn affected the decision-making process'. In other words, in order to understand the dynamics of policies related to Thailand's hydropower development, we must take into account the historical context and structural–agency relations in the analysis of particular policy outcomes.

Framed by a political ecology approach, the following sections look into two case studies in Thailand's hydropower development: the Pak Mun and Xayaburi dams. Using the two hydrological analogies of the 'backwater' and 'spillover' effects, this chapter argues that Thailand's natural resource management, especially the water-energy sector, has been part of the country's socio-politicized contestation and transformation. The backwater here refers to regulatory stagnation, to the repetition of the same old debates and public knowledge, as well as a lack of improvement in decision-making processes. The spillover, however, asserts that, within the same terrain of Thailand's ecological politics, the undercurrents of the backwater also create an unruly spillover effect on its neighbouring countries. While there has long been stagnation in hydropower development and policy improvement domestically, the complicated undercurrent of rising demands for power, on the one hand, and civil society's pressing concerns over the impacts of dams on the other, have forced the country to 'spill' its flawed hydropower decision-making processes over into neighbouring countries. Here, Thailand exports hydropower externalities to its neighbours so as to avoid domestic social and political pressure. The chapter now turns to look at the complex powers that underlie Thailand's hydropower decision-making processes.

The backwater: Pak Mun Dam

A quagmire of bureaucracy and knowledge politics

The Pak Mun Dam was built at the confluence of the Mun and Mekong Rivers in Thailand's Ubon Ratchathani Province. It was the first run-of-river dam in the Mekong Basin. With a length of 300m and a rated net head of 17m, the project's construction started in 1991 and its operation began in 1994. The Pak Mun Dam controversy is one of the best studied in the world and hence does not need rehearsing here (see e.g. Amornsakchai *et al.*, 2000; Roberts, 2001; Awakul and Ogunlana, 2002; Foran, 2006; Sneddon and Fox, 2008; Foran and Kanokwan, 2009; Jenkins *et al.*, 2008). Suffice to say, the key points of tension were:

- the poor consultation between the state and affected communities, over the dam itself, and over compensation and resettlement;
- the collapse of the fisheries in the area around the dam as a result of its construction;
- the failure of the government to deliver on promises to affected communities before and during dam construction;
- the bungled and confused efforts by the state to apply remedial measures that then failed.

The Pak Mun Dam controversy was so significant that it was rescaled into a wider debate about the need for domestic hydropower in Thailand. While the findings and recommendations from all of the research to some extent influenced policies and decision-making processes, the crucial undercurrent in the politics of Thai hydropower lies in stakeholder power relations and ineffective regulatory mechanisms, rather than in conflicting technical solutions (cf. Jakkrit, 2010). I will now turn to examine the governing mechanisms involved in the Pak Mun Dam, followed by the power relations among the stakeholders in their contestation over knowledge and solutions.

The case of the Pak Mun Dam has always been widely cited in public discussions. It has led to a debate over the local impacts on ecology – fisheries in particular – and the need for energy in the lower part of northeastern Thailand. The debates and knowledge constructed to support these arguments, however, seem to be perpetually stuck in political advocacy and disparaging remarks

Figure 5.2 Backwater vegetation covering the motionless reservoir behind the Pak Mun Dam

Source: author photo.

exchanged between state and civil society rather than leading towards long-term solutions or an improvement in decision-making processes. When looking back over the Pak Mun controversy, the knowledge produced to support the arguments and legitimize claims in the process of decision-making can be grouped into two broad categories. The first is the studies that support the construction of the dam, which claim that the EGAT conducted a technically legitimate operation within its own capacity and entitlement. The second category is the studies by civil society organizations and especially by locally affected people. The pro-Pak Mun Dam studies see water that flows from the Mun River into the Mekong as wasteful run-off that generates no economic value. This draws from a broader engineering-based narrative surrounding dam building and the availability of technologies in water resource development that materialized in the United States during the 1920s (Reisner, 1993). It is also, to some extent, the narrative that was used by the Mekong Committee in the 1970s–1980s. Within this narrative, building a dam at the confluence of the Mekong and Mun Rivers fundamentally yields benefits in terms of generating electricity and providing irrigation (Jakkrit, 2010). This perspective was later used to justify the decision to construct the dam (Wandee, 2000). Other justifications also included the propagation of freshwater aquaculture, the promotion of commercial agriculture, benefits for the Khong-Chi-Mun water diversion project, and the promotion of infrastructure development, transportation and tourism (EGAT, 2013).

The undercurrent and the emerging wave of new social movements

Knowledge production is part of the contested ground in hydropower decision-making (Jakkrit, 2010, 2012). In the case of the Pak Mun Dam, dominant knowledge produced by experts and decision-making authorities has been challenged by local studies articulated by villagers and their alliances with civil society. These counter-hegemonic studies have centred on the lack of opportunities for local people to participate in dam decision-making, and the impacts on their livelihoods. For many, the state's hydropower development policy has been seen as a threat, exploitation and transgression, which displaces people from their economic, cultural and social contexts. This has led to discontent that has prompted local people to conduct their own research to redefine the discourse and knowledge on the Pak Mun Dam's development (cf. Thai Baan Researchers *et al.*, 2002), which sets out to counter the state's claims to knowledge primacy.

Academics have conducted research into the issue and their findings confirm the protesters' statements. Some case studies that have been conducted compare the attitudes of local people and government officers towards environmental and resource development. This includes a case study on the conflicts between local people and state agencies (Decharat *et al.*, 1999). In this study, the changes in the socio-economic conditions after the construction of the Pak Mun Dam were examined. The study concludes the project yielded several negative impacts for the local community. The ecosystem was destroyed and this resulted in the decline of the fishery and loss of income. The abject failure of the dam's fish ladder provided ample fodder to those who resisted the dam (cf. Jutagate *et al.*, 2001, 2005; SEARIN, 2001). It is clear that the fish ladder technology did not allow fish to migrate upstream, and the explosion of islets and river rapids caused the destruction of the fish habitat and destroyed plant species.

These studies reveal the impact of the Pak Mun Dam on the river's ecology, and how local communities' attitudes and experience of the dam differ from those of the state. They reveal the lack of people's participation in decision-making, which finally led to the long-term conflicts over the dam's management. Within this politics of knowledge and policy advocacy, the state authorities have commissioned and employed research and technical know-how based on cost-benefit analysis, technical and managerial aspects of the project's operation, as well as the bureaucracy to justify the necessity of building hydropower projects. Local people too, with the help of their allies, have constructed their own knowledge to negotiate with the government for access to decision-making processes and rights to natural resource management. From the perspective of political ecology, the production of this knowledge is used strategically by each group of actors as a means of legitimizing their own positions and at the same time delegitimizing that of their opponents. In other words, rather than seeking comprehensive solutions to problems or creating platforms for inclusive decision-making, this knowledge production has created and intensified the polemics between the dam's supporters and its opponents. I argue that this quagmire of fragmented and disparate sets of knowledge has been one of the key challenges in the improvement of the hydropower decision-making process in Thailand and the Mekong Region. Dismantling the divide between opposing knowledge should be the starting point for constructive dialogues and participation involving all stakeholders (Jakkrit, 2013).

Previous studies and reports on the impacts of the Pak Mun Dam conducted by many institutions have lacked unity and have been uncoordinated (Kanokwan, 2004). This is due not only to methodological differences, but also to the political agenda behind them. Within this knowledge contention, protests have thus created a new form of political negotiation through the establishment of organizations, networks and movements, and the formation of alliances with NGOs, academics, local politicians and the media. Street protests have been a strategy deployed by local people to adjust the unequal power relations within hydropower politics and to wrestle decision-making from closed technocratic circles into the public arena. The social movement against the Pak Mun Dam started when General Chatichai Choonhavan (the then prime minister) approved the construction of the project. The villager protests resulted in the EGAT's approval to adjust periods for the opening of the dam's sluice gates each year. The assembly of social groups also empowered communities to oppose and challenge the state's power, as well as to turn the conflict into a national public issue (Somsri, 1999; Nalinee *et al.*, 2002; Missingham, 2003). Later, this strategy was utilized to build alliances that included not only people affected by the dam, but also other social groups who had a reason to oppose it. The state, however, also recreated knowledge and information with which to confront the protesters. For the state, it was the primacy of techno-science and national interest that mattered. Through cost-benefit analyses, feasibility studies and a variety of 'technical' research initiatives, the state not only tried to placate the protesters, but also attempted to appeal to a wider, national audience, who were doubtless alarmed by the state's claim that national electricity supplies were threatened. In turn, the protesters challenge these studies with their own studies, framed as 'citizen' or 'participatory' research, which focused on what they would lose (and did lose) as a consequence of the dam's construction.

The conflict over the Pak Mun Dam is one of the longest battles over natural resources between the state and local people in the history of Thailand's resource development. The enduring conflicts and the failure of the state to implement alternative decision-making processes have caused the country's hydropower development to stagnate. In this domestic context, the easiest option was to spill the effects of hydropower dam development into neighbouring countries where the undercurrents of civil society are not strong.

The spillover: Xayaburi Dam

The state's claim of increasing demand for electricity consumption and energy security in Thailand, while still debatable (Greacen and Greacen, 2004; Greacen and Palettu, 2007), is undeniably powerful in driving the agenda of dam construction forwards (Simpson, 2007). Dam construction, however, is only one among many contentious issues in the region's hydropower development. Others include water grabbing and allocation between sectors and countries (Matthews, 2012); resettlement planning and implementation; the sediment-trapping efficiencies of dams; the impacts of dams on fisheries and local livelihoods; inter-state tensions and hegemonies over water and its distribution, etc. In the case of Thailand, the domestic capacity for electricity generation alone, claims the EGAT, is insufficient for its household and industrial demands. Given robust social opposition to dam development at home, investments in hydropower production in neighbouring countries have emerged as political and short-term economically viable solutions. The Xayaburi Dam in Lao PDR is one among many spillover effects of the long-accumulated backwater of hydropower development in Thailand, that include the Yadana Gas Pipeline, Salween and Nam Theun 2 Dams (cf. Simpson, 2007, 2009). While it is not the first spillover effect of the domestic hydropower impasse, the Xayaburi Dam is, however, the first project to be built on the mainstream of an international river.

The spills of domestic backwaters

Among many projects that the Thai Government and companies are investing in in neighbouring countries is the Xayaburi Dam, the first hydropower dam ever built on the lower Mekong mainstream. The dam site is located in Xayaburi District in northern Laos, roughly 80 km south of the former capital, Luang Prabang. The dam is being constructed by Thailand's Ch. Karnchang Public Co. Ltd, with a value of US$3.8 billion. Once completed, the company claims that the dam will have an installed capacity of 1285 MW, and will generate 7405 GWh of electricity a year, supplying a million people in Laos and three million people in Thailand through EGAT's power grid (Pöyry PLC, 2012).

The Xayaburi Power Co. Ltd, which is a subsidiary of Ch. Karnchang PLC, acquired the project concession from the Lao Government. Ch. Karnchang PLC holds 30 per cent of the shares in the project. Unlike the Pak Mun, where funding came from international organizations like the World Bank, the Xayaburi Dam obtained loans from private financial institutions in Thailand, including the Siam Commercial Bank, Kasikorn Bank, Bangkok Bank and Krung Thai Bank (Save the Mekong, 2013). The shift in finance sourcing to commercial banks away from international development banks is occurring not only in Thailand but also across the Mekong Region (Middleton *et al.*, 2009). There are some crucial differences between private-sector investment and international development bank investment that make it a challenge for civil society to demand accountability from dam builders. First of all, most Thai commercial banks do not have clear responsible business and corporate social responsibility policies governing the provision of loans to transnational mega-projects. Also, in the case of the Xayaburi, with a coalition of loans from at least four banks, it is easy for the lender to shirk their responsibilities and accountability when impacts occur, especially outside their own territory. The shift from international development bank to commercial bank investment is thus one of the key transitions in allowing the backwater to endure.

While the financial sources and construction management differ between the Pak Mun and the Xayaburi, the fundamental design of the dams is similar: the run-of-the-river dam or the 'transparent dam' according to the Lao authority (see Chapter 6 in this volume). The run-of-the-river structure of the Xayaburi Dam, it is often claimed, was designed to avoid creating a large reservoir. Xayaburi will be built as a reinforced concrete dam 820 m long with a rated net head of 18 m. The dam will be equipped with one 60 MW turbine generating electricity for Lao PDR; and seven 175 MW turbines generating electricity most of which will be exported to Thailand. According to the design, the dam will also feature a navigation lock and fish passage facilities to allow the migration of fish and other aquatic animals through or around the dam structure (cf. Baumann and Stevanella, 2012). Additional emergency spillway gates have been arranged to cope with seasonal water highs and to control flooding. Theoretically, the reservoir head level will be kept largely constant and inflowing water will constantly be released either through the turbines or the spillway, so as not to store the water and to maintain a mean daily flow all year round. When construction is completed, it is expected that the project will generate electricity for 29 years, beginning in 2019, under a contract based on the build-operate-transfer principle.

The new watershed in Thailand's hydropower decision-making

Because the Xayaburi is on the Mekong mainstream, it is the first dam to fall under the MRC's PNPCA process. The rationale for construction of this dam is explained by the memorandums of understanding signed by the governments of Laos, Cambodia, Myanmar and China to promote electricity generation by member countries for sale to Thailand (EPPO, 2013). The dam's developers claim that Xayaburi Dam will have a positive effect on the livelihoods of the Lao people, state welfare services, developing the education system and creating employment (Xayaburi Power, 2013). The dam is expected to yield considerable income and other economic benefits for Laos (Stone, 2011) that will help to achieve its poverty elimination target by 2020.

These, then, are the dominant narratives of the Xayaburi Dam (see also Chapter 6 in this volume). The counter-narratives focus on damage to the transnational ecology of aquatic ecosystems and livelihoods along the river. According to the studies by Costanza et al. (2011), ICEM (2010), Baran et al. (2011) and others, the effects will be seen at both the local and regional levels and will affect a large number of people in the Mekong Basin due to the destruction of aquatic habitat, fisheries and other ecological resources, not to mention claims that the dam will trap sediment, thereby affecting the Tonle Sap Great Lake in Cambodia and Vietnam's Mekong Delta (cf. Stone, 2011).

With regard to the impacts on people and their livelihoods, it is estimated that the construction will force more than 2,100 people in the project area to relocate. Additionally, more than 200,000 farmers and fishermen in four nearby districts in Lao PDR will experience negative impacts on their livelihoods due to their dependence on natural resources for food and economic security (International Rivers, 2011). James Leape, the Director General of WWF International, has proposed that, if the Lao Government completes the Xayaburi Dam, it will threaten opportunities for economic development, as well as the food security of millions of people in the region (Leape, 2013). These people depend, both directly and indirectly, on the resources in the river for their incomes and food. The natural flow of the Mekong River, it is argued, nurtures agricultural activities worth an estimated US$4.6 billion (Leape, 2013). And while our understanding of the Mekong ecosystem and approximate fish catch is far from complete (Grumbine et al., 2012), the value of fisheries in the area where the impacts from the dam can be felt is hard to anticipate.

Several concerns have been raised by civil society groups regarding the decision-making surrounding the Xayaburi (Baran et al., 2011). First of all, the project lacks transparency and disclosure of the full information from the developer's EIA (Pöyry PLC, 2012). It is argued that the Xayaburi construction received approval before proper research was conducted both on its potential impact on the river system as a whole, and also in relation to additional Mekong dam construction projects. According to strategic environmental assessment of mainstream Mekong dams commissioned by the MRC, construction projects on the Mekong River will cause grave negative effects on ecology, economy and society, both domestically and transnationally. It is, therefore, necessary to disclose this information to the public, particularly to those people who will suffer a direct impact (ICEM, 2010). In addition, there are doubts about the quality of the EIA reports conducted by the dam's developers. For example, the WWF Greater Mekong's study, which analysed the 2008 feasibility study and the 2010 EIA report, commented that the 'studies do not meet the standard expected for a sustainable hydropower project in reference to current best practice elsewhere in the world' (Baran *et al.*, 2011, 4).

In sum, the Xayaburi Dam can be seen as one of the key spillover effects of Thailand's backwater in its domestic decision-making impasse. Developed to exploit rich natural resources in Lao PDR with direct foreign investment from Thailand, the project is considered as the most immediate and viable solution to the stagnation of hydropower politics in Thailand. Rapid economic development will significantly increase regional electricity demands, which ultimately makes the construction of the Xayaburi Dam an attractive and politically viable option. Building the dam in Lao PDR will allow dam developers to avoid resistance from the Thai civil society organizations where their political and social campaigns are limited within the Thai territory. The crucial challenges of hydropower development here are: whether the decision-making process regarding this dam was carried out with transparency and accountability; and whether the Lao Government and Thailand-based developers, rather than the impacted people in Lao PDR and riparian area, have more power to manipulate resources for their own benefits. If the project allowed a more inclusive participatory process that could benefit the decision-making, this would create better alternatives and choices for management strategies that would probably entail more sustainable results than the present approach.

Conclusion: the changing current in Thailand's hydropower politics

An examination of the Pak Mun and Xayaburi dams has shown how Thailand's hydropower decision-making has been shaped by civil society, transnational cooperation and economic regionalization. The controversial case of the Pak Mun has been a key turning point for hydropower development in Thailand. The local impacts and the contention between local villagers and the state authority have, for the past few decades, expanded into stagnant hydropower decision-making at the national level. In addition, Pak Mun contributes very little to the Thai national grid (TDRI, 2000), and this should have prompted Thai decision-makers to reflect and think through how they could avoid similar situations when developing later projects. Key leaders of social movements in the country who I have interviewed argue that the Thai Government keeps the Pak Mun in place so as not to lose face by accepting the failure of the dam and hence delegitimizing its decisions to commission and operate future dams. The politics of image keeping here is a prime obstacle for the state in allowing greater people's participation and trying to find pragmatic solutions in managing the dam. Such stagnation has created what I call a 'backwater', where knowledge, narratives and debates over solutions to domestic hydropower development have reached an impasse. As Thailand's production capacity is unable to meet the country's energy demands, investing in and purchasing electricity from its neighbouring countries has consequently become an attractive alternative. The Xayaburi project is fundamentally driven by such a need. It is also driven very substantially by the construction profits that are reaped at an early stage.

The domestic backwater, however, is not a situation created by the state authorities alone. Thai civil society organizations have also contributed to the impasse in the improvement of hydropower decision-making processes and the spillover effect. When asked about the fact that Thailand is now exporting the issue to its neighbouring countries, most of the Thai activists I interviewed only blamed this problem on the energy-producing sectors. But are the domestic social movements not part of this equation? Do they not contribute to the backwater as well? The anecdotes of the street and river protests that I described at the beginning of this chapter and the reluctance of civil society organizations to work concomitantly with the state and private sectors have undeniably created an atmosphere that obstructs constructive engagement in improving decision-making processes. By positioning themselves simply as an anti-dam movement, civil society organizations have often been viewed as a blockade that has created a deadlock to hydropower development in the country. The spillover of development projects to neighbouring countries is thus a result of animosity that has been maintained by both the state and domestic civil society organizations.

While Thailand can overcome its backwater limitation, the spillover of hydropower projects to its neighbours will have some impacts on local people and ecology. Such effects, if not well prepared for and managed, could lead to basin-wide impacts in which a regional mechanism is needed to resolve any contention. With transboundary challenges, regional decision-making processes in developing and managing hydropower dams across the border become inescapable. In order to improve hydropower development at both the domestic and regional levels, and to ensure social, economic and ecological sustainability, there are some crucial points that need to be taken into account in the decision-making processes.

First of all, there is a need for stakeholders to participate in the development of hydropower projects from their initiation. Pak Mun's contention, to a certain extent, derives from the lack of participation of the people in the project's initiation and implementation. But the Pak Mun case was only domestically bounded. With the spillover, participation processes will need to embrace stakeholders across international borders in order to legitimately capture the wide-ranging impacts beyond the national territory. In addition, participation should not be seen merely as a form of 'public relations' by dam investors, construction companies and state agencies advocating for the dam. From our Pak Mun lesson, it seems that multi-stakeholder decision-making processes happened too late when actors were on the cusp of negotiating the finance and the concession agreement. Having meaningful participation of all stakeholders at the early stages and, in the case of Xayaburi, across borders, will reduce complicated contentions and negotiations.

In addition, past antagonistic sentiments should be carefully managed and avoided in any future discussion over the issue of new dam development. One of the lessons learned from the Pak Mun case is that past contentions have led to the development of entrenched animosity between opposing sides, creating significant obstacles to finding solutions together. In fact, positive sentiments among stakeholders are important not only for improving the decision-making processes, but also for the acceptance of diverse knowledge and cultural legitimacy. We have learned that the politics of knowledge derogation can create and exacerbate contention rather than enabling problem-solving. Multiple approaches and cooperation among the various actors in conducting research together should enable cross-sectoral understanding and the generation of solutions. Knowledge justification in the decision-making process should not be limited to so-called 'experts' in dam and energy development, but should take other actors who accumulate and articulate distinctive kinds of knowledge into consideration (Jakkrit, 2012). The contrast of scientific and technical knowledge with local knowledge should be dismantled and different actors be allowed to claim legitimacy in the decision-making process based on their own epistemologies.

Lastly, the new phase in Thailand's hydropower development involves multiple states and transnational corporations. The social responsibility and accountability required by national laws and local social pressure will be insufficient when hydropower development becomes a transboundary issue. On the one hand, civil society organizations and affected people should create a region-wide alliance in order to campaign, voice their concerns, and engage in an international regulating mechanism for regional hydropower development. On the other hand, regional and basin management should not be exclusive to state-based cooperation and negotiation, but should also involve actors from civil society across the region. The PNPCA process, for example, should allow civil society organizations and affected people to take part and play an active role in the consultation and regional decision-making. Without recognized channels, civil society and the affected people will have to take to the streets and rivers to express their grievances.

In order to move environmental politics back into the mainstream, there is a need to improve the negotiation infrastructure and regulatory mechanisms to include a wide range of stakeholders in the decision-making circle. Like a healthy river encompassing a diversity of fish, flowing sediments and living vegetation, healthy decision-making processes in hydropower development should also include multiple actors in its political ecology dynamics. Simply spilling backwaters into neighbouring countries will not solve ecological politics. Rather, its effects will create a greater swamp of impacts that will cause the deterioration of the whole basin in the long run.

References

Amornsakchai, S., Annez, P., Vongvisessomjai, S., Choowaew, S., Thailand Development Research Institute (TDRI), Kunurat, P., Nippanon, J., Schouten, R., Sripapatrprasite, P., Vaddhanaphuti, C., Vidthayanon, C., Wirojanagud, W., and Watana, E. (2000) *Pak Mun Dam, Mekong River Basin, Thailand: A WCD Case Study Prepared as an Input to the World Commission on Dams*, Cape Town: Secretariat of the World Commission on Dams.

Awakul, P., and Ogunlana, S. O. (2002) 'The effect of attitudinal differences on interface conflict on large construction projects: The case of the Pak Mun Dam project', *Environmental Impact Assessment Review*, 22(4): 311–35.

Baran, E., Larinier, M., Ziv, G., and Marmulla, G. (2011) *Review of the Fish and Fisheries Aspects in the Feasibility Study and Environmental Impact Assessment of the Proposed Xayaburi Dam on the Mekong Mainstream*, Gland, Switzerland: World Wildlife Fund Greater Mekong Subregion Program.

Baumann, P., and Stevanella, G. (2012) 'Fish passage principles to be considered for medium and large dams: The case study of a fish passage concept for a hydroelectric power project on the Mekong Mainstem in Laos', *Ecological Engineering*, 48: 79–85.

Bryant, R. (1992) 'Political ecology: An emerging research agenda in third-world studies', *Political Geography*, 11(1): 12–36.

Bryant, R. (1999) 'A political ecology for developing countries? Progress and paradox in the evolution of a research field', *Zeitschrift fur Wirtschaftsgeographie*, 43(3–4): 148–57.

Chenaphun, A. (2012) 'Laos holds groundbreaking ceremony for contentious Mekong Dam', Reuters, US edition, 7 Nov., www.reuters.com/article/2012(11)/07/us-laos-dam-idUSBRE8A618I20121107, accessed Aug. 2013.

Costanza, R., Kubiszewski, I., Paquet, R., King, J., Halimi, S., Sanguanngoi, H., Luong Bach, N., Frankel, R., Ganaseni, J., Intralawan, A., and Morell, D. (2011) *Planning Approaches for Water Resources Development in the Lower Mekong Basin*, Chiang Rai: Portland State University and Mae Fah Luang University.

CUSRI (Chulalongkorn University Social Research Institute) and ERI (Environmental Research Institute Chulalongkorn University) (2001) *Process of People Participation in the Analysis of Environmental Impacts*, Bangkok: Chulalongkorn University, Office of Natural Resources and Environmental Policy and Planning (in Thai).

Decharat, S., and Young Generation Group for the Studies of Alternative Economy (1999) *The Change in Economic and Social Conditions After the Construction of Pak Mun Dam: The Study of Biological Diversity and Agricultural Systems around Pak Mun Dam Area*, Ubon Ratchathani: Ubon Ratchathani University (in Thai).

EGAT (Electricity Generating Authority of Thailand) (2013) 'The benefits of Pak Mun Dam', www.egat.or.th/thai/pakmoon/pakmoon4.html, accessed Jan. 2013 (in Thai).

EPPO (Energy Policy and Planning Office), Ministry of Energy, Thailand (2013) 'Xayaburi Project', www.eppo.go.th/power/xayaburi-pp.pdf, accessed Jan. 2013.

Foran, T. (2006) 'Rivers of contention: Pak Mun Dam, electricity planning, and state–society relations in Thailand, 1932–2004', PhD thesis, University of Sydney.

Foran, T., and Kanokwan, M. (2009) 'Pak Mun Dam: Perpetually contested?', in F. Molle, T. Foran, and M. Käkönen (eds), *Contested Waterscapes in the Mekong Region: Hydropower, Livelihoods and Governance*, London: Earthscan, pp. 55–80.

Greacen, C., and Palettu, A. (2007) 'Electricity sector planning and hydropower in the Mekong Region', in L. Lebel, J. Dore, R. Daniel, and Y. S. Koma (eds), *Democratizing Water Governance in the Mekong Region*, Chiang Mai: Mekong Press, pp. 93–125.

Greacen, C. S., and Greacen, C. (2004) 'Thailand's electricity reform: Privatization of benefits and socialization of costs and risks', *Pacific Affairs*, 77(3): 517–41.

Grumbine, R. E., Dore, J., and Xu, J. (2012) 'Mekong hydropower: Drivers of change and governance challenges', *Frontiers in Ecology and the Environment*, 10(2): 91–8.

Hirsch, P. (1999) 'Beyond the nation state: Natural resource conflict and national interest in Mekong hydropower development', *Golden Gate University Law Review*, 29(3): 399–414, www.internationalrivers.org/files/attached-files/the_xayaburi_dam_-_thai_final.pdf, accessed Jan. 2013 (in Thai); www.savethemekong.org/news_detail.php?nid=180&langss=th, accessed Jan. 2013.

ICEM (International Center for Environmental Management) (2010) *MRC Strategic Environmental Assessment of Hydropower on the Mekong Mainstream: Summary of the Final Report*, Vientiane: Mekong River Commission.

International Rivers (2011) 'Xayaburi Dam: A looming threat to the Mekong River' www.internationalrivers.org/files/attached-files/the_xayaburi_dam_eng.pdf, accessed Jan. 2013.

Jakkrit, S. (2010) 'Hydraulics of power and knowledge: Water management in northeastern Thailand and the Mekong Region', PhD thesis, Australian National University.

Jakkrit, S. (2012) 'Decolonizing the river: Transnationalism and the flows of knowledge on Mekong ecology', in L. S. Kook, S. Huang, and M. Hayes (eds), *Revisiting Transnationalism in East Asia: Emerging Issues, Evolving Concepts*, Seoul: Jimoondang, pp. 175–87.

Jakkrit, S. (2013) 'Representing community: Water project proposal and tactical knowledge', in R. Daniel, L. Lebel, and K. Manorom (eds), *Governing the Mekong: Engaging in the Politics of Knowledge*, Selangor: Strategic Information and Research Development Centre, pp. 11–26.

Jenkins, K., McGauhey, L., and Mills, W. (2008) *Voices from the Margin: Economic, Social and Cultural Rights in Northeast Thailand: Pak Mun Dam*, Khon Kaen: Peace and Human Rights Center of Northeast Thailand.

Jutagate, T., Lamkom, T., Sataporanwanit, K., Naiwinit, W., and Petchuay, C. (2001) 'Fish species diversity and ichthyomass in Pak Mun reservoir, five years after impoundment', *Asian Fisheries Science*, 14: 417–24.

Jutagate, T., Krudpan, C., Ngamsnae, P., Lamkom, T., and Payooha, K. (2005) 'Changes in the fish catches during a trial opening of sluice gates on a run-of-the-river reservoir in Thailand', *Fisheries Management and Ecology*, 12: 57–62.

Kanokwan, M. (2004) 'The politics of negotiation for the opening of Pak Mun sluice gates', *Matichon Newspaper*, 23 June, p. 7 (in Thai).

Leape, J. (2013) 'In the Mekong, science – not guesswork – must prevail', World Wildlife Fund International, wwf.panda.org/who_we_are/wwf_offices/laos/newsrom/?206911/In-the-Mekong-science-not-guesswork-must-prevail, accessed Jan. 2013.

Lebel, L., Dore, J., Daniel, J., and Koma, Y. S. (eds) (2007) *Democratizing Water Governance in the Mekong Region*, Chiang Mai: Mekong Press.

Matthews, N. (2012) 'Water grabbing on the Mekong Basin: An analysis of the winners and losers of Thailand's hydropower development in Lao PDR', *Water Alternatives*, 5(2): 392–411.

Middleton, C., Garcia, J., and Foran, T. (2009) 'Old and new hydropower players in the Mekong Region: Agendas and strategies', in F. Molle, T. Foran, and M. Käkönen (eds), *Contested Waterscapes in the Mekong Region: Hydropower, Livelihoods and Governance*, London: Earthscan, pp. 23–54.

Missingham, B. (2003) *The Assembly of the Poor in Thailand: From Local Struggles to National Protest Movement*, Chiang Mai: Silkworm Books.

Molle, F., Foran, T., and Käkönen, M. (eds) (2009) *Contested Waterscapes in the Mekong Region: Hydropower, Livelihoods and Governance*, London: Earthscan.

Nalinee, T., Sulaiporn, C., and Siriporn, K. (2002) 'The experience of struggle by Mun riparian people: The cases of the Pak Mun and Rasi Salai Dam', in P. Pasuk (ed.), *Livelihoods and Struggle: Contemporary People's Movements*, Bangkok: Trasvin (in Thai).

Pöyry Public Limited Company (2012) 'Xayaburi Hydro Power Plant, Lao PDR: Background Material on Pöyry's Assignment', 9 Nov., www.poyry.com/sites/default/files/imce/eng_xayaburi_hpp_09112012_final.pdf, accessed Jan. 2013.

Reisner, M. (1993) *Cadillac Desert: The American West and its Disappearing Water*, revised and updated edn, New York: Penguin Books.

Roberts, T. R. (2001) 'On the river of no returns: Thailand's Pak Mun and its fish ladder', *Natural Historical Bulletin of the Siamese Society*, 49: 189–230.

Save the Mekong (2013) 'Chor Karnchang expected to sign the contract early next year', www.savethemekong.org/news_detail.php?nid=180&langss=th, accessed Jan. 2013 (in Thai).

SEARIN (Southeast Asia Rivers Network) (2001) 'The lessons from the decommissioning of dams worldwide', www.livingriversiam.org/4river-tran/others/wd_a2.html, accessed Jan. 2013 (in Thai).

Simpson, A. (2007) 'The environment–energy security nexus: Critical analysis of an energy "love triangle" in Southeast Asia', *Third World Quarterly*, 28(3): 539–54.

Simpson, A. (2009) 'Transnational energy projects and green politics in Thailand and Burma: Critical approach to activism and security', PhD thesis, University of Adelaide.

Sneddon, C., and Fox, C. (2008) 'Struggles over dams as struggles for justice: The World Commission on Dams (WCD) and anti-dam campaigns in Thailand and Mozambique', *Society and Natural Resources*, 21(7): 625–40.

Somsri, C. (1999) *Pak Mun Dam: The Origin of Conflict in Thai Society, 1987–1995*, Ubon Ratchathani: Ubon Ratchathani University (in Thai).

Stone, R. (2011) 'Mayhem on the Mekong', *Science*, 333(6044): 814–18.

TDRI (Thailand Development Research Institute) (2000) *Report for the World Commission on Dams Pak Mun Case Study*, Bangkok: TDRI.

Thai Baan Researchers, Assembly of the Poor, and SEARIN (Southeast Asia Rivers Network) (2002) *Thai Baan Research on Mun River and the Return of Fishermen: Summary and Knowledge of Fishes of Pak Mun People*, Bangkok: SEARIN (in Thai).

Thavivongse, S. (1998) *EIA: The Analysis of Environmental Impact Assessment*, Bangkok: Green World Foundation (in Thai).

Wandee, S. (2000) 'Ten years of Pak Mun Dam: The struggle of the poor fighters', *Feature Magazine*, 184 (June): 82–107 (in Thai).

Xayaburi Power (2013) 'Xayaburi Power Company Limited', www.xayaburi.com/pnpca.html, accessed Jan. 2013.

6 The invisible dam

Hydropower and its narration in the Lao People's Democratic Republic

Kim Geheb, Niki West and Nathanial Matthews

> The Xayaboury dam is a state-of-the-art dam; it is a kind of transparent dam, which means it is a dam without a dam.
>
> (Aloune Xayavong, Deputy Secretary-General of the
> Lao National Mekong Committee, 23 November 2012)

Introduction

Of Laos's 26,500 MW hydropower potential, 18,000 MW are considered technically exploitable. Almost all of this potential resides in the Mekong and its tributaries. This well-watered nation contributes about 35 per cent of the Mekong's annual discharge of 475 km^3, and is well suited to hydropower development. Its highlands rise to 2,800 m (the Annamite and Luang Prabang ranges, and the Bolevan and the Xiangkhoang plateaux); and lowlands typically are below 500 m. It is a country of copious forest, large water resources, significant mineral deposits and – for now at any rate – very few people (its population in 2012 was an estimated 6.5 million – just 26.7 people per km^2).

In this chapter, we analyse the narratives and contested space surrounding hydropower development in Laos, before focusing on the Xayaburi Dam. Narratives surrounding hydropower often delineate the spaces between the winners and losers of development. Narratives are carefully constructed by agents to support agendas and delegitimize critics (Roe, 1991). In this chapter, we critically analyse the highly contested narratives surrounding hydropower to illuminate the broader political, economic and historical forces that precipitate development in Laos. This is not necessarily simple. Narratives, while an attractive focus of analysis, are not normally clear-cut, depending on who spins them, and for what purposes. Our analysis of narratives and the policy statements that support them demonstrates some of the reasons why Laos has not only refused to back down from its ambitious and controversial development of the mainstream, and how it attempts to portray its hydropower ambitions more generally.

There are two lead narratives in the Mekong. The first is that delivered by the Lao state, which portrays the rapid development of hydropower as a solution to the nation's poverty and least developed country status, and which serves as a means to attracting foreign direct investment (FDI). Hydropower will yield significant revenues that can be used in society's best interest. 'Complementary narratives' (possibly developed for different reasons and to serve different interests) from international agencies echo this sentiment, especially the World Bank, the Asian Development Bank and a variety of Chinese, Thai, Malaysian and other developers and financiers. The second narrative is that of international non-governmental organizations (INGOs), whose narrative focuses on environmental destruction (especially of fisheries), and human rights (especially of resettled communities), and which have served to problematize hydropower development. Other narratives complement these, particularly those of increasingly strident regional NGOs and civil society interests; and of the scientific community, again possibly for different reasons and/or interests.

Laos's decision to embark on a mainstream dam – the Xayaburi – accentuated this contrast, and drew in many more observers than the country had previously needed to confront. Narratives emerging from Laos during this contest assumed metaphysical dimensions: that the dam was so benign, it would be as if it was not there. We trace this debate closely, and analyse it from the perspective of narrative formation and use.

Narrative formation in Laos

For Harvey (1993, 25),

> all ecological projects (and arguments) are simultaneously political-economic projects (and arguments) and vice versa. Ecological arguments are never socially neutral any more than socio-political arguments are ecologically neutral. Looking more closely at the way ecology and politics interrelate then becomes imperative if we are to get a better handle on how to approach environmental/ecological questions.

Narratives emerge as mythologies around particular constructed problems. If we are to accede to Harvey's perspectives, these mythologies emerge to furnish ends, usually of those who created the narrative in the first place. The reason why they are perceived as myths is because they are unanchored in science; this, in turn, necessitates their construction for they have no foundation (cf. Forsyth, 2003). In Laos, their inspiration is often derived from abroad, equating problems from elsewhere with perceived problems in their country.

Perhaps the best example of a Laotian narrative is the shifting cultivation narrative. Shifting cultivation has long been characterized as a destructive farming system. Starting in the 1980s, Laos began to embrace this externally derived narrative.

> Swidden agriculture has only become apparent as a problem within the last 20–30 years as increasing population and decreasing primary forest areas

have sparked a global awareness that swidden agriculture is no longer a viable, sustainable technology in most areas. Laos is no exception, although in most provinces the downward spiral of environmental degradation leading to an ever-decreasing carrying capacity in the face of ever-increasing populations – this deadly spiral has just begun.

(Defour, 1994, 229)

The Lao Government portrays 'slash-and-burn' as 'a symbol of backwardness and an absolute environmental evil' (Aubertin, 2001, 11). 'These ethnic groups have lived on slash-and-burn rice cultivation which dries up streams, ravages green forest, leaves mountains bare' (KPL News Agency, 4 July 1990, quoted in Ireson and Ireson, 1991, 929). Such statements serve to 'catastrophize' (the art of giving weight to the worst possible outcome, no matter how unlikely it is) the narrative, suggesting urgency and magnitude, and therefore to legitimize government actions that seek to address it. The shifting cultivation narrative in Laos, indeed, has inspired a raft of legislation, policy, legal instruments and the creation of a wide variety of bureaucracies (cf. Rigg, 2005).

This 'downward spiral' is an important characteristic of the narrative, and as always, is beguilingly simple: shifting cultivation causes deforestation; run-off therefore increases, along with soil erosion; this then has hydrological and sediment implications; flooding and drought therefore increase. Land use in upland areas therefore threatens livelihoods in lowland areas. This duality, Aubertin (2001) argues, is necessary to sustain the dominant swidden agriculture narrative. There exists very little research-based evidence, however, to confirm the narrative (cf. Lestrelin, 2010; Lestrelin *et al.*, 2012). Shifting cultivators are mainly non-Lao ethnic groups; by demonizing shifting cultivation, such ethnic groups are also demonized: 'upland populations are sometimes denounced as "dangerously backward and ignorant"' (Lestrelin, 2010, 434).[1] For many observers, the Lao Government's uplands and land use policies are intended to obscure a resource grab, seeking to wrest control of valuable forestry and hydropower resources from ethnic communities (Lestrelin, 2010; Baird and Shoemaker, 2007). Laos's Focal Zone Development Strategy, which is closely related to their various uplands agricultural policies, seeks to resettle shifting cultivators in lowland villages, intending to sedentarize them, and therefore make it easier to provide them with education, sanitation, health and other services (cf. Bartlett, 2012). The state frequently portrays such resettlement as a privilege for those communities that supported the revolution. In a recent article on a resettlement initiative in the country's southern Sekong Province, 'the relocation is considered an obligation of the party and the government for local heroes who contributed to the country's revolution' (*Pasaxon*, 2014). Such relocation villages are also intended to integrate these communities into the market economies practised by the lowland Lao, thus making them 'more Lao' (Rigg, 2005, 97; Ireson and Ireson, 1991). In addition, the shifting cultivation narrative serves to draw attention away from more obvious deforestation sources. In Laos, the military controls the country's logging activities (Lestrelin,

2010; Evans, 2002; Hodgdon, 2008); in 2006, forest products were thought to account for 12 per cent of government revenue (Barney and Canby, 2011). But Laos is not alone in its preoccupations with shifting cultivation. In what might be called a 'discourse coalition' (Forsyth, 2005), Laos's concerns for shifting cultivation dovetail with those of Western donor agencies and NGOs, for whom environmental, conservation and community development concerns are *raisons d'être*.

Narratives, Hajer (1995, 64–5) writes 'are devices through which actors are positioned, and through which specific ideas of "blame" and "responsibility" and "urgency" and "responsible behaviour" are attributed'. Narratives form in Laos in much the same way as they do in other countries: they involve a myth, unsupported by science; they serve to reinforce the existing power and/or claims of a dominant actor over other less powerful actors; they serve to legitimize; they serve to catastrophize; and, indeed, they serve to apportion blame. It is these building blocks of narrative formation that we shall now trace as we consider a brief history of the dialectics surrounding Laos's initial forays into hydropower. By analysing the political and economic forces behind pro-hydropower narratives put forward by the government of Laos (GoL) this chapter begins to illuminate why the discourse surrounding hydropower development in Laos between the government and INGOs is so polarized.

The first narrative: developing the basin

Following the 1954 Geneva Conference, during which Cambodia, Laos and Vietnam all obtained independence from France, the United Nations Economic Commission for Asia and the Far East (ECAFE) prepared a report titled the *Development of Water Resources in the Lower Mekong Basin*. The report called for 90,000 km² of irrigation to be developed, the construction of five dams to generate 13.7 GW to be delivered through an interconnected grid. Based largely on the ECAFE recommendations, the Committee for Coordination of Investigations of the Lower Mekong Basin (more conveniently, 'the Mekong Committee') was established in September 1957. The Committee itself comprised representation from four of the lower Mekong's countries, Vietnam, Laos, Thailand and Cambodia, and initially had a focus that was specific to flooding, to be implemented through the creation of a Flood Control Bureau. The Mekong is a river overwhelmingly characterized by floods. Of the 475 km³ of water that it deposits into the South China Sea annually, 75 per cent is delivered during the monsoon, between July and October. The Mekong, in other words, cannot but flood. And its populations have adjusted to these floods, as the quintessential image of a 'Mekong house' will testify: it is on stilts. But a focus on floods provided a very real opening for the Mekong Committee. By portraying floods as dangerous, costly and damaging, it proclaimed the need to *control* the river, rendering it amenable to further development (cf. Wong, 2010).

While the Mekong Committee received funding from a variety of international sources, principal support came from the US, who were keen to see the Mekong Region developed in the hopes that this would thwart communist advances in the region (Molle *et al.*, 2009). This would be regional pacification via engineering, as the Mekong Committee's *Indicative Basin Plan* (Mekong Committee, 1970) showed. Inter alia, it called for 30,000 km² of irrigation by the year 2000, 87 short-term tributary development projects and 17 long-term development projects on the Mekong mainstream. In total, 180 possible projects were identified. The 'poster child' (as Molle *et al.*, 2009, refer to it) for this initiative was the Pa Mong Dam, proposed for the Mekong mainstream just west of Vientiane. The dam would be 98 m high, would store over 100 billion m³ of water, have an installed capacity of 4000 MW, irrigate some 2 million hectares and inundate a total area of almost 4000 km² across both Laos and Thailand. It would displace 250,000 people, a figure that was later revised upwards to 400,000. At the cost of US$1 billion, the Pa Mong was to be the world's largest multi-purpose dam at the time (Molle *et al.*, 2009). Growing insecurity in Laos, however, forced the Pa Mong plan to be shelved. Nevertheless, the Pa Mong plans revealed the magnitude of the developments that the Mekong Committee had in mind.

By 1978, just 16 of the proposed 180 projects had been implemented. This included the Nam Ngum 1 north of Vientiane, which was completed in 1971. Its construction had been completed under difficult conditions. Its site lay in territory controlled by the Pathet Lao, Laos's anti-royalist, pro-communist forces, who, while they attacked USAID and American military barracks stationed close to the dam, carefully left the dam itself and Canadian construction workers alone (Thi, 1999). This first phase of the dam had cost US$28 million with contributions from the US[2] (almost 50 per cent of budget), Thailand, Japan, the Netherlands and others. Like most dams that would later be built in Laos, a significant proportion of the dam's electricity would be exported to Thailand, at the time the region's primary energy market. Usefully, two islands in the dam's reservoir were put to use as re-education camps for prostitutes and drug addicts after the 1975 overthrow of the Lao kingdom (Stuart-Fox, 2013).

Following the rise of the Khmer Rouge in Cambodia, the Committee was short of a member, and was disbanded as a result in 1978 (see Chapter 1). Instead, an Interim Mekong Committee was established, which issued, in 1980, a Revised Indicative Basin Plan, the aspirations of which were somewhat more modest than the first plan. It called for the development of a cascade of smaller dams along the Mekong's mainstream, divided into 29 hydropower and irrigation projects, three of which were international in scope. Cambodia, under the Khmer Rouge, was excluded from the plan.

These Indicative Basin Plans have tremendous resonance in the Mekong's development. They painted a future of immense developments, of engineering able to command the forces of nature to benefit humanity. The Indicative Basin Plans did *not* paint a picture of integrated energy systems supplied by thousands of small-scale energy plants interspersed with great tracts of forest pregnant with biodiversity. The vivid maps in these documents dotted the entire basin with one dam after the next, their reservoirs redolent in indigo hues. These spoke not of subduing the river, but its complete enslavement. Despite their comprehensiveness, the plans are simple: dams can generate energy, which can in turn be sold for money, in turn yielding funds that pave the way to modernity. The Indicative Basin Plans were road maps to modernity, emblemized by physical infrastructure, but connoting sophistication, the predominance of science and material comfort (not least electric lighting and air conditioning). For Wong (2010), the Mekong Committee created a space 'amenable to development', but in many respects, it did more than that: it showed the way.

The Indicative Basin Plans continue to have visible influence. The successor of the Interim Mekong Committee, the Mekong River Commission (MRC), has its own 'Basin Development Plan Program', which traces its history explicitly back to these plans (MRC, 2013).

In 2006, the World Bank and the Asian Development Bank (ADB) collaborated with the MRC to produce the Mekong Water Resources Assistance Strategy (MWRAS) (ADB and World Bank, 2006). Just 10 per cent of the Mekong Basin's hydropower potential was being utilized, the MRWAS said, and argued that the

> Mekong basin has flexibility and tolerance, which suggests that sustainable, integrated management and development can lead to widespread benefits. This may contrast with the more precautionary approach of the past decade that tended to avoid any risk associated with development, at the expense of stifling investments.
>
> (ADB and World Bank, 2006, 4)

The report is replete with references to 'sustainability', 'dialogue', 'Integrated Water Resources Management' and 'balanced development', but unashamedly privileges hydropower investment and water development at the expense of alternative uses for water, or other methods of generating electricity; it also completely ignores the warnings captured in the Report of the World Commission on Dams (WCD, 2000; see also Matthews, 2013).

The MWRAS was one of a variety of documents and measures the World Bank was putting in place to announce its return to hydropower development after a ten-year hiatus (in part arising from the concerns raised in the WCD Report: Matthews, 2013). But the MWRAS dovetailed neatly with the Lao state's own internal ambitions. Laos is a single-party state. At its helm is the Lao People's Revolutionary Party (LPRP), central to which is the ambition of 'finishing the revolution' through the rapid restructuring of the Lao society and its economy (Bartlett, 2012). New partnerships with foreign investors have enabled it to accelerate this process in what has been referred to as the LPRP's 'big push'.

> The scale and speed of the investment in dams, mines and plantations are a big push attempt to rapidly lift Laos out of its low-income situation. The government's publicly stated goal is that by 2020 Laos will have graduated

from the league of least developed countries. Rapid development of natural resources on the scale now taking place in Laos holds the promise of generating substantial, ongoing revenues that might be recycled into significant increases in public spending (another aspect of big push strategy) to address shortcomings in agriculture, food production, education, healthcare, infrastructure and so on.

(Fullbrook, 2010, 10)

Laos is not alone in the use of 'big push' strategies, and experience from elsewhere in the world suggest that these strategies are prone to significant lags between actual developments and the institutional capacity needed to maintain and manage them (Fullbrook, 2010; Bartlett, 2012). But the Lao state may feel that the BOOT formula (see below) ensures that such maladjustment between infrastructure and institutional capabilities is irrelevant: given that private companies own and operate dams in Laos for the duration of their concession, Laos has time to catch up with its own institutional strengthening before dams are handed over.

The 'big push' pervades Laos's Seventh Five-Year National Socio-Economic Development Plan (2011–15) (MPI, 2011). The word 'rapid' appears 26 times in the plan; while the term 'mega-project' appears 11 times. Mega-projects are seen to comprise mining, industrial plantations and hydropower, which feature prominently. The construction of eight hydropower plants is envisaged within the plan period, with a combined installed capacity of 2862 MW. The plan then goes on to identify 26 hydropower projects by name, as well as stating the intention to 'carry out a study and thereafter construct hydropower projects along the Mekong' in Luang Prabang Province (MPI, 2011, 192).

Laos's five-year socio-economic development plans are the state's central narratives; all other narratives emerge from these documents. Because the LPRP 'is the people', and 'is the state', the five-year plans express the will of the Lao people, and there can be no justification greater than that (cf. Stuart-Fox, 2005). They also, of course, serve to perpetuate the state in its current form (cf. Bruce St John, 2006).

The second narrative: a dam fracas

By the late 1980s, Soviet aid to Laos was dwindling. Laos, a nation that had always relied very heavily on international aid, was in trouble. As a result, the LPRP announced its 'New Economic Mechanism' in 1988, which closely followed changes in other communist countries: glasnost in the Soviet Union, *doi moi* in Vietnam, and the new reforms in China. These all suggested that the trappings of capitalism could be obtained while still preserving the country's state apparatus (Evans, 2002; Bruce St John, 2006).

In turn, these moves allowed it to engage with the international development banks (IDBs), and a raft of legislation followed that sought to encourage private enterprise, direct foreign investment, property rights, etc. The IDBs, for their part, stressed the need for Laos to take advantage of its natural resources, of which hydropower was seen to have significant potential (hence, the MWRAS). The IDBs proposed that Laos venture into the hydropower arena by drawing on so-called 'buy–own–operate–transfer' (BOOT) formulas, under which private funding is used to develop the facility, which is then developed, managed and operated as a private enterprise for a specified period, before being transferred to the state (the 'rent-a-river' formula: Usher and Ryder, 1997) (see Chapter 7 in this volume).

The Nam Theun-Nam Kading is a Mekong tributary in eastern-central Laos. The Interim Mekong Committee's Revised Indicative Basin Plan identifies three sites along the river's course which it considered particularly well suited to hydropower, especially the Nam Theun 2 (Wong, 2010). In the event, it was another dam called the Theun-Hinboun that was to be built first, the first attempt to implement the BOOT strategy in Laos. The Theun-Hinboun Power Company (THPC) was established in 1994 through a shareholder agreement between the Lao Government, Thailand's MDX Lao Public Company and Nordic Hydropower AB. The formation of THPC was significant for Laos at the time, as a symbol of new economic policy and cooperation with neighbouring countries and private investors. Funding for the dam (about US$240 million) was raised through loans from the ADB and the Nordic Development Fund. Additional funding was raised through commercial banks and export credit agencies. Theun-Hinboun marked the first major venture under the new foreign investment policies and the first time the government had formally engaged the private sector to build a power plant. The government nominated Electricité du Laos (EdL) as holder of its 60 per cent interest in the project, with the other two shareholders each taking a 20 per cent stake. Construction started in November 1994, and was completed ahead of schedule and under budget by March 1998 (THPC, 2014).

The ADB and other international donors played a key role in legitimizing the THPC; central to their arguments were claims relating to engagement with the market, rather than more nuanced environmental and developmental arguments. The ADB itself described its investment in the project as 'aimed at enhancing foreign exchange earnings through export of electric power from the Lao People's Democratic Republic to Thailand' (ADB, 2013). The Western origins of these arguments presented ideological difficulties for Laos, but none eclipsed the country's revenue problems.

Because of the THPC's focus on electricity supply to Thailand, its planners had completely neglected environmental and social safeguards. An environmental impact assessment (EIA), funded by the Norwegian Agency for Development Cooperation (NORAD), and carried out by a Norwegian firm, NORCONSULT, was widely criticized by international NGOs and in Norway (a country involved in the project in multiple ways: Usher and Ryder, 1997). NORAD followed up in a variety of ways, including commissioning the Association for International Water and Forest Studies (FIVAS), a Norwegian NGO, to carry out a study to investigate whether or not the concerns being aired had any validity. FIVAS's report described multiple problems, including absent consultations with project-affected people prior to the start of construction; the absence of any formal compensation policy; and the underestimation of social and environmental costs (FIVAS, 1996). These types of concerns have become *de rigueur* for INGO criticisms of Lao hydropower development, and form a central part of their own narratives.

Doing neither itself nor the project any favours, the ADB wrote in 1997 that the bank could 'derive satisfaction from having backed a winner', and that 'there is little for the environmental lobby to criticize in Theun-Hinboun' (ADB, 1997, 8). On the contrary, there was plenty that the environmental lobby could focus on in THPC's planning and implementation. In 1998, an International Rivers report followed up on the FIVAS work, repeating similar criticisms and raising similar concerns: villagers throughout the project area and downstream were suffering greatly from the impacts of the project, lost livelihoods and were not being adequately compensated, if at all (Shoemaker, 1998).

In the end, more than 30 environmental groups from around the world sought to criticize, influence or stop the project. Never in Laos's history had anything similar been encountered. In particular, the sustained pressure from International Rivers eventually affected both the ADB and, more profoundly, THPC, which developed a significant (and costly) social and environment programme in an effort to deflect some of the criticism (cf. Whitington, 2012).

The THPC experience established Laos as a key battleground between hydropower development and the environmental lobby. Narratives that privileged the market and electricity supply over and above environmental and social concerns were strongly challenged. Importantly, for the environmental lobby, THPC, while resistant to many of their arguments, was also vulnerable to them, not least because of the variety of transparency and accountability rules that it had to comply with, which in turn provided the INGOs with the fodder they needed to contest the dam.

These transparency and accountability rules – perhaps because of the THPC's experience – were largely absent from subsequent dam projects in the country. Very little opposition was levelled at the Houay Ho Dam in southern Laos (commissioned in 1999), the Nam Leuk (2000), the Nam Mang 3 (2004), or the Xeset 2 (2009), if for no other reason than that there was very little information available about them (this despite the ADB being involved in several of these projects).

The Nam Theun 2 (NT2) project was to be, however, a whole other kettle of fish. Once again, the ADB and the Nordic Development Fund (among others) were involved in this initiative; but so too, after its decade-long absence from hydropower, was the World Bank. The developers' narrative was strongly influenced by the World Bank's own strategic positioning. The NT2 would demonstrate that 'sustainable hydropower' was no oxymoron; on the contrary, the unlikely marriage of environment and development was possible, as the Brundtland Report had argued (WCED, 1987). This strongly articulated argument from the World Bank meant that they had to 'get it right' on the NT2. It 'is intended to exemplify the Bank's achievement of socially and environmentally responsible development. If the idea of World Bank-directed development is not supported by its flagship project in Laos, then its applicability elsewhere becomes highly questionable' (Singh, 2009, 488). NT2 was to be a 'kinder, gentler dam' (to quote *Newsweek* in 2007[3]) because it would meet the World Bank's and the ADB's environmental and social safeguards. These too called for varying degrees of transparency and accountability,[4] with the result that the project drew massive INGO criticism and, needless to say, attracted immense scrutiny as a consequence, better described elsewhere (Lawrence, 2009; Singh, 2009; Mirumachi and Torriti, 2012; Hirsch, 2002). In order to sustain the idea of a 'clean, green dam', the World Bank set about creating and/ or overhauling much of Laos's environmental and social legislation, in a process that Goldman (2001) calls the construction of the 'environmental state'. Laos's Science, Technology and Environment Administration (STEA) emerged from this process, and laid the foundation for today's Ministry of Natural Resources and the Environment (MONRE). Presently, virtually all the Laotian hydropower policy, legislation, and associated institutions were created or strengthened during the implementation of the NT2 project.

The NT2 was commissioned in 2010. Other dams followed: the Nam Lik 2 (2010), the Nam Ngum 2 (2010), the Nam Ngum 5 (2011) and the Xe Kaman 1 (2011). While there was certainly INGO attention to these dams, nothing approached the clamour that the NT2 had generated. The latter was attacked because it could be attacked; ADB and World Bank safeguards and policies demanded that the project be scrutinized at every turn. To gain information about these other dams, many INGOs had to resort to occasional and furtive visits to their sites; documentation about these dams was patchy at best and non-existent at worst; EIAs were rarely released and were superficial if they were; and dam development was frequently commenced before an EIA had been prepared.

For Laos, the basis for much of the INGO protest was uncomfortable, for it called for significant adjustments to how the state as a whole conducted itself. In many respects, Laos still held on to the future portrayed in the Indicative Basin Plan; and its own five-year plans, ones that argued that the gains from hydropower and massive engineering by far and away eclipsed environmental and social concerns. In any case, it has already achieved 'sustainable hydropower' through the THPC and NT2 projects (cf. Thoummavongsa and Bounsou, 2013), and the dense, World Bank-supported, legal and regulatory framework from which sustainable hydropower is supposed to emerge.

The environmental and social safeguards deployed for these two dams, while not passing muster for most regional and international NGOs, are exceptional. The 'National Policy on Environmental and Social Sustainability of the Hydropower Sector in Lao PDR', explains that the NT2 has 'been inseparably linked to the development of national legal and institutional provisions for environmental and social safeguards' (Lao People's Democratic Republic, 2006, 11). NT2, the policy notes, has 'defined new environmental and social standards for a development project in Lao PDR' (Lao People's Democratic Republic,

2006). But these points may be moot. As Viraphonh Vilavong, the Lao Vice Minister for Energy and Mines put it:

> [y]es, they are saying that Nam Theun 2 is a very good project. But to use it as a standard, it's not possible. We can use it as a good example, a good guideline, but not as a standard. All the developers say that it is not possible to use Nam Theun 2 as a standard.
>
> (quoted in Jusi, 2011, 256)

Lao Government officials frequently remark that if they were to fully apply their own hydropower policy, guidelines and legislation, no one would ever invest in the sector. The ADB also recognizes that Laos finds its own safeguard policies trying, 'which governments [in the region] increasingly see as a nuisance' (quoted in Matthews, 2013, 150).

A duality has emerged. On the one hand, Lao hydropower policy and legislation demonstrate that the country is committed to sustainable hydropower; and the THPC and the NT2 prove its commitment. On the other hand, tens of additional dams were being built, largely unaffected by the state's own environmental and social safeguards policy and legislation. NT2 and THPC are unique dams in Laos, not just because of the ADB and World Bank inspired safeguards that they employ. They are the only two hydropower companies in Laos, which, from start to finish and thereafter had European shareholders,[5] for whom reputational risk is a significant factor in their business calculations. Arguably, the social and environmental performance of these two dams has little to do with Lao legislation and policy, and everything to do with the World Bank's and the ADB's involvement in their development, and the presence of Western shareholders in their ownership.

In this sense, a narrative developed and proffered by the ADB and the World Bank has helped Laos obscure other, less careful, hydropower investments. The narrative that governs the latter draws almost entirely from the Indicative Basin Plan, and sits uneasily with the World Bank/ADB narrative. Nevertheless, it was the World Bank/ADB narrative that drew the ire of the INGOs' counter-narrative that characterized the IDBs and the hydropower developers (and not necessarily the Lao state) as bent on profiteering, 'green washing', supressing dissent and lack of transparency.

The third narrative: the invisible dam

In 2010, the MRC released an influential document, the Strategic Environmental Assessment (SEA) of Hydropower on the Mekong Mainstream, written by the International Centre for Environmental Management (ICEM, 2010). ICEM argued that, although Cambodia and Laos would benefit from the FDI, royalties, revenues and domestic electricity supply (Laos would receive 70 per cent of the economic benefits) of mainstream hydropower, the costs to fisheries, agriculture, biodiversity and ecosystem integrity would outweigh these benefits, and inequality and poverty in the basin would increase in the short to medium term. Moreover, institutions to implement and enforce regional standards and safeguards were not, they said, fully developed; nor did Mekong countries have the institutional or human resources capacity to effectively plan, manage, monitor, oversee and regulate the construction and operation of mainstream dams without incurring significant social and environmental costs. After consultation with multiple stakeholders in MRC member countries, the SEA recommended that mainstream hydropower development be delayed for ten years, to give countries and researchers time to gather baseline data, conduct further impact assessments and explore less destructive and more innovative options (ICEM, 2010).

The Mekong SEA was one of several reports that saw the entrance into the Mekong dams debate of what, for many observers, was a more credible variety of science than that commissioned by NGOs or dam developers. ICEM was a regionally respected institution; the SEA was released by the MRC, which served to legitimize it, and the methods employed in the SEA attracted no criticisms from the regional and international scientific community. The SEA was the first in a succession of research outputs that had begun to query dominant pro-hydropower 'all benefits, no costs' narratives in the basin (cf. Costanza *et al.*, 2011; Ziv *et al.*, 2012). Perhaps most importantly, these studies were conducted at scale, developing scenarios for the system as a whole, rather than the relatively local scales of previous analyses.

The SEA was important to priming the debate that was about to unfold, for it explored a future that had already begun to emerge in the Mekong Committee's Indicative Basin Plan, where Xayaburi is clearly identified as a potential mainstream dam site. In 2007, the GoL signed a Memorandum of Understanding (MoU) with Ch. Karnchang, a Thai firm who already had experience building dams in Laos with the Nam Ngum 2, to develop the Xayaburi, the first mainstream dam in the lower basin.

The MRC has no regulatory authority in the affairs of the Mekong save one: if any of its member states intends to build a dam on the mainstream of the Mekong, then it must comply with the MRC's Procedures for Notification, Prior Consultation and Agreement (PNPCA). The purpose of the PNPCA is for riparian countries to discuss the potential transboundary impacts of the proposed project, and, ideally, to reach consensus on how development should proceed. Ultimately, however, the Mekong Agreement does not supersede national sovereignty, and a country may still proceed with the proposed development without obtaining approval from the other countries under the 1995 Mekong Agreement.

The GoL had, as explained above, remained largely silent in the debates around the NT2 and the THPC. It had, nevertheless, been involved in the construction of multiple other dams which it had carefully kept from public scrutiny. Because Laos is a signatory to the 1995 Agreement that created the MRC, it now had to comply with the PNPCA process, and GoL had to show its cards. This included sharing the dam's EIA, in which the dam is not referred to as a dam at all, but as a 'barrage' 'with a river pond', because of its run-of-river design (Matthews, 2013). As a result, the EIA argues, the dam will not affect fisheries production but, on the contrary, 'will improve the overall natural fisheries production capacity' (quoted in Matthews, 2013, 268). It seems probable that it is in this framing of the dam by the EIA that the narrative on its invisibility emerged. So too, the simplicity of these narratives was retained: because the Xayaburi is run-of-river, ergo no damage to fisheries production would occur.

The Xayaburi Dam triggered the MRC's first experience with the PNPCA process, which officially began in October 2010. The MRC Secretariat, in conjunction with member countries, began reviewing Xayaburi's technical documents to determine its compliance with the MRC's guidelines on environmental and social impacts, particularly those of a transboundary nature. Among the documents submitted for review were the feasibility study and social and Environmental Impact Assessments (EIAs). Countries downstream of Laos had hitherto remained largely silent on Laos's dam building activities, but the fact that the Xayaburi was on the Mekong mainstream had significant potency (even though it is by no means clear that the cumulative impact of many mainstream dams will be greater than that of tributary dams (see Ziv *et al.*, 2012)). Both Vietnam and Cambodia were to enter into the Xayaburi debate robustly. Broadly speaking, Cambodia's primary concern was for the fisheries of the Toné Sap Lake; while for Vietnam, it was the dam's sediment trapping efficiency and how this might impact the Mekong Delta.

Because of the PNPCA process, these documents were placed in the public domain, where criticism was scathing. The scientific community decried the EIA's claim that the Xayaburi would have minimal effects on fish migration, sediment transport and flows due to its run-of-river design; furthermore, the EIA only contemplates potential impacts up to 10 km downstream of the dam, a limited study area. Large and credible research institutions waded into the debate. ICEM reviewed the EIA, and stated that, despite the dam's run-of-river design, it was still likely to have significant impacts on fish migration, sedimentation and flow regulation (Vaidyanathan, 2011). The World Fish Centre was commissioned by the World Wildlife Fund (WWF) to review the EIA, and found it equally flawed (Baran *et al.*, 2011).

One month after Xayaburi entered the prior consultation process, Laos signed a concession agreement with Xayaburi Power Company Limited (XPC), a subsidiary of Ch. Karnchang PCL, in which it holds 57.5 per cent of shares. Four Thai banks financed 70 per cent of the project, with Ch. Karnchang financing the remaining 30 per cent. The Electricity Generating Authority of Thailand (EGAT) will purchase 95 per cent of the electricity generated. The GoL stands to earn roughly $4 billion over the 29-year concession period, through the collection of royalties and taxes.

The PNPCA process was carried out over the course of six months, from late 2010 until April 2011. At the end of October, the MRC began its technical review of the dam's design and its proposed impacts on fish migration and sedimentation. The public consultation period commenced in January 2011, and MRC member countries – with the exception of Laos – held public consultations throughout January and February 2012. The final draft of the

MRC's prior consultation report was released at the end of February (MRC Secretariat, 2011) and an MRC Joint Committee meeting was held on 19 April to discuss the report's findings and reach an agreement on the Xayaburi project proposal.

After reviewing the MRC's final prior consultation report, Thailand, Vietnam and Cambodia wanted to extend the prior consultation process due to concerns that: transboundary and cumulative impacts had not been adequately addressed; gaps in the data and studies remained; potential impacts on the environment and livelihoods were not clear; and more public consultations were needed. The three countries were adamant that the six-month prior consultation process was insufficient to thoroughly evaluate a project of the Xayaburi's magnitude. Vietnam went as far as stating that no mainstream dams should be built for ten years, in accordance with the SEA's recommendation. Laos argued that extending the prior consultation process was unnecessary and impractical, as transboundary impacts would be unlikely, any additional studies would take longer than the six-month prior consultation process, and it had already complied with the PNPCA's stipulations. Unable to reach a consensus at the 19 April meeting, the MRC member countries agreed that the prior consultation process would need to be continued at the ministerial level.

Irrespective of this process, however, newspaper reports a few days prior to the meeting indicated that Laos has been proceeding with preparatory work at the Xayaburi site 'on the sly' (*Bangkok Post*, 2011). Internationally and regionally – particularly in Cambodia and Vietnam – newspaper reportage was strident and sceptical of the GoL's position. The *New York Times* quoted government responses to the MRC report, among them that the Xayaburi's impact would be equivalent to a 'natural waterfall', and that hydroelectricity is 'green energy [and] shall be strongly promoted and supported' (*New York Times*, 2011).

These assertions of the neutral impact of the dam were to become even more intriguing. The Lao Vice Minister of Energy and Mines, Viraphonh Vilavong explained in a 2011 interview that the Xayaburi 'is a run-of-river scheme. It means that the input flow is the same as the output flow. It's like having no dam there. So this is considered a transparent dam' (Varchol, 2012). Also of interest was persistent reference to 'international' experts in the regional media, as well as by GoL:

> We are using international standards for the development of the Xayaburi Hydropower Project. We have acquired international consultants with a lot of experience, even though it is a little expensive. And these are things that should comfort the people and the neighbouring countries that we are doing this in a very transparent way.
>
> (Viraphonh Vilavong, interviewed in Varchol, 2012)

The international consultants referred to were the Swiss branch of a Finnish consulting firm, Pöyry, who were hired to assess whether or not the Xayaburi Dam's design met the MRC's Design Guidelines. It should be noted that

Pöyry was also under consideration at the time by the Lao Government to be Xayaburi's lead engineering firm, which International Rivers considered a potential conflict of interest.

At the May 2011 ASEAN Summit, the Lao Prime Minister told his Vietnamese counterpart that Laos would suspend construction of the Xayaburi Dam (*Thanh Nien News*, 2011). Similar reassurances were delivered to then US Secretary of State Hillary Clinton (*Time*, 2011). The US had voiced concerns over the dam's potential impacts, particularly on regional stability (*Phnom Penh Post*, 2011a).

Laos did not, however, suspend work on the Xayaburi Dam site. In June, a letter from the Department of Energy Promotion and Development at the Lao Ministry of Energy and Mines was leaked to the press (*Phnom Penh Post*, 2011b). The letter, dated 8 June and addressed to the Xayaburi Power Company, claimed that Pöyry had completed its study and found that the Xayaburi Dam was in compliance with MRC's Design Guidelines, thus 'the prior consultation of the Xayaburi project has now been completed and the prior consultation process has ended at the MRC Joint Committee level' and 'we hereby confirm that any necessary step in relation to the 1995 Mekong Agreement has been duly taken in a spirit of cooperation and working together of all relevant parties' (*Phnom Penh Post*, 2011b). The latter made no reference to the continued ministerial consultations that were supposed to have followed the 19 April meeting. In reaction to the letter, XPC sent a letter to the governor of EGAT, repeating the claim that Laos had met its obligations under the Mekong Agreement, so a power purchase agreement could now be signed (*Phnom Penh Post*, 2011b). Discussion in the media focused on why Laos had even bothered to embark on the PNPCA process when it appeared increasingly to be heading towards unilateral action. In response, Vice Minister Viraphonh was quoted in *Time Magazine* as saying 'If you need a specific agreement before you can do something, nothing happens … Whether it's good or bad, I don't know, but there is no development' (*Time*, 2011).

The Pöyry compliance report was to attract very significant criticism. International Rivers, the WWF and ICEM all released critiques of it (ICEM, 2011; Herbertson, 2011; WWF, 2011). They noted that, while Pöyry did identify inadequacies in the dam's technical documents, it failed to understand the complexities of the Mekong River system, and made inappropriate recommendations with respect to fish passage, dam break analysis, sediment routing and fish stocking (ICEM, 2011). The report also stated that the prior consultation process was completed at the 19 April MRC Joint Committee meeting. The report's claim that 'the Xayaburi HPP has principally been designed in accordance with the applicable MRC Design Guidelines' (WWF, 2011) drew considerable ire. International Rivers argued that the dam could not be in compliance if it recommended over 40 additional studies be conducted (Herbertson, 2011). In addition, the fact that the report's executive summary reached conclusions that differed from those in its technical write-up was targeted. The technical portion of the report acknowledged that additional baseline data were needed, and that the EIA did not sufficiently address sedimentation, erosion, water quality, fisheries and dam safety issues (ICEM, 2011). In their view, recommending that studies and design improvements take place during the construction period, without first acquiring the baseline data noted as absent, was proof of Pöyry's 'build now, adapt later' (ICEM, 2011) approach and neglect of the precautionary principle, which therefore violated the MRC's Design Guidelines (WWF, 2011).

Following release of Pöyry's report, Xayaburi PCL signed a power purchase agreement with EGAT (*Thanh Nien News*, 2012). Subsequently, Cambodia, Vietnam and the MRC Secretariat conducted their own reviews of the report, concluding that the report did not provide justification of compliance with the MRC's Design Guidelines. An MRC ministerial-level meeting was held in December 2011 and this time countries managed to agree that construction of the Xayaburi Dam should be suspended until further studies addressing the dam's impacts could be conducted (*Asian Scientist*, 2011).

GoL responded by commissioning French dam developer, Compagnie Nationale du Rhône (CNR), to peer-review Pöyry's report, during which time activity continued unabated at the dam site. In April, Ch. Karnchang signed a contract with the Xayaburi PCL to build the Xayaburi Dam, and submitted it to the Thai Stock Exchange, with the announcement that construction on the dam had begun on 15 March (*The Diplomat*, 2012).

At the end of March, CNR completed its review. While this desk study provided neither new insights, nor covered fishery impacts or migration, it was used as a basis to call for a 'redesign'. Vice Minister Viraphonh was quoted as saying that 'this study … confirms that if the Lao government wants to let the dam be redesigned, there will be no impact on the environment' (RFA, 2012a). The redesign would cost, it was said, US$100 million (RFA, 2012a). These statements suggested a shift in GoL thinking. Although it implied that the previous dam design would have affected the environment, a 'redesign' would assure its environmental benevolence. An already benign dam had become even more benign.

April and May saw an escalation in demonstrations against the Xayaburi Dam. Notably, these demonstrations (mainly in Thailand) were being carried out by regional citizens. Cambodia took a more aggressive stance and threatened to sue Laos in international court: 'There must be a discussion before Laos can proceed with the construction. If Laos has decided unilaterally, then according to law, we can file a complaint to an international court', said the Permanent Vice Chairman of Cambodia's National Mekong Committee (UPI, 2012) In May, Cambodia's representative to the MRC delivered a formal letter to his Lao counterpart asking that Laos respect the agreement made at the MRC's December 2011 meeting: 'Cambodia's position is that Laos should halt the dam construction while the environmental impact study is being carried out' (RFA, 2012b). Vietnamese scientists joined the fray and encouraged the Vietnamese Prime Minister to issue a formal complaint as well (RFA, 2012a). The Network of Thai People in Eight Mekong Provinces protested outside the Mekong2Rio conference in Phuket. The network's leader threatened to shut down the

Lao–Thai Friendship Bridge if the MRC did not act to stop the construction of the Xayaburi Dam (*Nation*, 2012). The meeting was attended by senior government officials from the region, including Vice Minister Viraphonh who reiterated that Laos would not proceed without consent from the other MRC member countries, and that work to date was only for site preparation. 'We will address and take into account all reasonable concerns in order to make this Xayaburi Dam a transparent dam and a role model for other dams in the mainstream of the Mekong River. Even the turbine to generate electricity was a fish friendly version' (ANN, 2012). This was the first time that a Lao official had publicly stated the country's intention to develop additional dams on the Mekong mainstream.

In June, International Rivers released pictures from a trip to the Xayaburi Dam site. The pictures depicted dredging of the river and construction of a concrete retaining wall. International Rivers showed the pictures to the MRC and ICEM. The MRC's comment was: 'Our engineering-based observation is that the reported activities show that the preparation for construction is well-advanced. However, we are not certain if/why the dredging is taking place, nor why the concrete wall is being built' (*Thanh Nien News*, 2012). ICEM's response was more blunt: 'It's quite clear that the company is just constructing the dam full pelt ahead. There is some definite infill and platforms being built in the mainstream just offshore. So they've gone from being purely access road construction to putting in foundation walls and offshore pads for the dredging machinery to work off' (BBC, 2012a).

At the mid-July ASEAN ministerial meeting, after another month of protest and negative press, Laos's Minister of Foreign Affairs announced that work on the Xayaburi Dam would, once again, be suspended so that additional studies could be completed (*The Economist*, 2012). A week later, however, GoL announced that the government would not interfere with Ch. Karnchang's implementation plans, including village resettlement (SEAWATER, 2012). 'An amazing week', wrote *The Economist*, 'of conflicting statements, stark contradictions and confusion has made everything about the site of a controversial dam project at Xayaburi, in northern Laos, as clear as mud' (*The Economist*, 2012). Japan also reported that Laos had not yet approached it to support a new EIA, which it had agreed to do at the 8 December MRC ministerial-level meeting (Reuters, 2012a).

In the same month, GoL hosted a tour of the dam site for international donors, ambassadors, MRC officials and NGOs, who received presentations from Pöyry and CNR, stating that additional studies were under way, that sedimentation problems had been solved (WWF, 2012), and that 85 per cent of the fish would be able to safely pass through the dam (*The Economist*, 2012). For participants on the tour, it seemed clear that work was proceeding as planned and on schedule.

As work at the dam site continued, protesters maintained a strong resistance. In August, a group of villagers from Thailand's Mekong villages filed a lawsuit in the Thai Administrative Court against EGAT, based on a constitutional provision that allows citizens to sue enterprises – including state-owned enterprises – that undertake environmentally harmful projects without adequate public consultation or impact assessment (*Eco-Business*, 2012). NGOs in Finland submitted a complaint against Pöyry to the government, alleging ethical misconduct and a violation of the OECD Guidelines for Multinational Enterprises (*DW*, 2012). The Secretary-General of the Cambodian National Mekong Committee was vocal in reiterating Cambodia's opposition to the dam at an ASEAN meeting in September: 'We have made comments about the potential transboundary impacts. [Laos] really did not respond to those concerns. They remain unanswered' (*Phnom Penh Post*, 2012).

In November, James Leap, the International Director General of the WWF, wrote a blog piece titled 'In the Mekong, science – not guesswork – must prevail' (Leap, 2012). In it, he wrote, 'The fish that migrate up and down the free-flowing lower Mekong are the principal source of protein for those 60 million people, and are the basis for a fishing industry with an estimated value as high as $7.6 billion annually. And the river's natural flooding cycles feed agriculture that brings in another $4.6 billion. So the stakes are high.' It was an excellent example of catastrophication: the dam *could* affect fish stocks; *could* affect livelihoods; and *could* affect agriculture throughout the *whole* system.

In that same month, Laos hosted an event of which it was immensely proud: the Asia-Europe Summit Meeting (ASEM). It was during this that the country suddenly announced that it had held a groundbreaking ceremony for the Xayaburi Dam. The CEO of Ch. Karchang maintained that the dam would not have any untoward environmental impacts: 'If [it would] badly affect the environment, we wouldn't do it. This company wouldn't do it. This is the company's strongest policy' (Reuters, 2012b). Vice Minister Viraphonh concurred: 'I am very confident that we will not have any adverse impacts on the Mekong River. But any development will have changes. We have to balance between the benefits and the costs' (BBC, 2012b). This too suggested a change in position. The dam was now not as benign as previously.

The Deputy Prime Minister of Laos claimed that the MRC countries had finally reached an agreement: 'We had the opportunity to listen to the views and opinions of different countries along the river. We have come to an agreement and chose today to be the first day to begin the project' (Reuters, 2012b). The Thai Government issued a statement at ASEM, saying that it was not opposed to the Xayaburi Dam (*Bangkok Post*, 2012). Cambodia and Vietnam did not react publicly, expressing neither support nor opposition. Vice Minister Viraphonh was certain that Laos had finally won them over: 'We can sense that Vietnam and Cambodia now understand how we have addressed their concerns. We did address this properly with openness and put all our engineers at their disposal. We are convinced we are developing a very good dam' (BBC, 2012b). Following ASEM, Pöyry was awarded the engineering contract for the Xayaburi, and the GoL officially approved the project in its National Assembly.

Cambodia and Vietnam had the opportunity to air their grievances in the New Year at the MRC Council meeting on 15–17 January. Vietnam asked that no more work on the Xayaburi take place until a four-year study could be finished (IUCN, 2013). Cambodia said that Laos must have misunderstood the 1995 Mekong Agreement, as the prior consultation process was never completed (International Rivers, 2013). MRC donors from Australia, Japan, Europe and the United States issued a joint statement expressing their continued concern about the Xayaburi Dam's impacts: 'It is our consensus that building dams on the mainstream of the Mekong may irrevocably change the river and hence constitute a challenge for food security, sustainable development, and biodiversity conservation' (Herbertson, 2013). Laos declined to sign the meeting minutes.

In many respects, neither Cambodia nor Vietnam could afford to push the fight much further without threatening their own interests. Vietnam's arguments were difficult to sustain, given that Vietnam's Central Highland dams already trap very significant amounts of sediment from this rich sediment source (Kummu *et al.*, 2010; Kummu and Varis, 2007); and at the time of the PNPCA, Cambodia was close to signing off on the Lower Sesan 2 Dam, which is implicated in very significant fisheries impacts (Ziv *et al.*, 2012) (see also Chapters 7 and 8 in this volume). In addition, Cambodia is also seriously considering its own mainstream dams (Sambor and Stung Treng). There are wider political concerns too: Vietnam is keen to maintain its historically close relationship with Laos so as to counter the spread of Chinese influence in the southeast Asian peninsula.

Conclusion

By the time of the 1975 revolution, Laos had only one large hydropower dam (i.e. of 15 MW installed capacity and above); by 2012, it had 12 additions, with 12 more under construction. The base narrative for all of these projects talks of the country's need to generate electricity for development, and to generate the income needed to lift the country out of 'least developed country' status by 2020.

The Nam Ngum 1 Dam presented few difficulties for Laos to implement. Its implementation occurred under conditions of civil strife, and in a time where the rhetoric of the Mekong Committee provided sufficient weight to obscure any dissenting voices: at this time, the imperative of development was sufficient to ensure that opposing views would be treated as madness.

The echoes of this logic have endured, as the MWRAS showed, the MRC's focus on Basin Development Plans, and the Lao state's own five-year socio-economic development plans all emphasize and privilege the idea that there are direct relationships between large, rapidly implemented, capital investments and significant social and economic benefits.

Emerging from its socialist experiment, Laos bought into ideas proposed by Western development banks; it had little choice but to do so. The country was on the brink of bankruptcy. Initially, with the THPC, Laos was presented with the BOOT model, which must have seemed immensely attractive to the cash-strapped country. The cost (and skill) of building dams was deferred to the developer who, in turn, assumed the risk of the investment. Whatever furore erupted between the ADB and the environmental lobby was not Laos's to address.

The same was true of the NT2. Here, the vexation was between the INGOs and a World Bank keen to ensure that environment and development were perceived as two sides to the same coin. In this, the Lao state did not involve itself, and turned instead to applying the BOOT model with other, less environmentally and socially discriminating investors. This included early engagements with Chinese, Korean and Thai developers; and, indeed, saw the entrance of Chinese capital in the country's hydropower sector.

The World Bank's parallel investments in Laos's environmental policy and institutions, and the existence of THPC and NT2 as 'sustainable' hydropower investments provided Laos with environmental credentials that it used to obscure less careful hydropower investments. Given that the Xayaburi had been on the books for some time, and Laos knew that it would have to submit them to the MRC's PNPCA process, it is possible that it imagined that these credentials would help to deflect some of the criticisms that would inevitably come. Using these narratives to shroud its real intent was a strategy Laos had already employed in its management of the shifting cultivation 'issue'.

And come the criticisms did. Those offered by the INGOs were to be expected; but new actors entered the scene, including downstream countries; regional NGOs and civil society movements; independent scientific interests; international governments; and immense interest from regional and international media. Laos had never come under such immense scrutiny that buried what it considered a perfectly valid narrative to justify the dam: income with which to develop the country. It interpreted the outcry as preoccupied with the environment, more particularly, the fisheries and, to a lesser extent, sediments. As a result, Laos moved to demonstrate that the dam would be environmentally benign; so much so, that it would be as if it were not there at all. It 'redesigned' the dam at great cost to incorporate fish- and sediment-friendly technologies. While at times clumsy and contradictory, Laos stayed the course. As the debate over the environmental consequences of the dam swirled, other more practical issues needed solution, not least the PNPCA process. Effectively, Laos simply withdrew from the process, thereby leaving the MRC battered and weakened. While it had run the PNPCA process effectively and correctly, Laos's engagement in the institution is central to its legitimacy. Through the conduit it represented, however, Laos was obliged to face criticisms levelled at it by other member countries, which did not – and still do not – accord with its development aspirations.

The spring has been reloaded and a new debate awaits. In the south of the country, another dam, the Don Sahong, is planned for a channel in the Four Thousand Islands (Si Pan Don) area. Critics argue that this channel is the deepest of the Si Pan Don channels, and is therefore used by many more migratory fish than the other channels. This is not a problem, the dam's developers have replied, saying that they will excavate and deepen a nearby channel to accommodate the fish migration. But the dam will affect downstream Irrawaddy dolphin populations in Cambodia at Kratie the critics have said. This is not the case, the developers and the government have replied. Cambodia and Vietnam have, once again, have been vociferous in their complaints.

Accusation, rebuttal and counter-rebuttal has recommenced, but the Lao state has grounds, now, to be more confident of itself. It was able to submit the Xayaburi plans and impact assessments to the MRC, receive significant comment and criticism, and still proceed with the construction of the dam, not via consensus, but through unilateral action. The Don Sahong is, technically, on the Mekong mainstream. Laos notified the MRC of its intentions just a month before construction on the dam was planned to start, arguing that the dam was not subject to a normal PNPCA process because it did not span the entire breadth of the Mekong mainstream. In the case of the Xayaburi, issues were obscured behind an argument that the dam was so benevolent that it was as if the dam was not there; in the emerging Don Sahong narrative, the dam is not really a mainstream dam. It seems highly likely, buoyed with the success of the Xayaburi, that Laos will continue to use obfuscation to legitimize their claims over the hydropower of the Mekong mainstream; and in turn, to use these major 'narrative dramas' to obscure continued damming of Mekong tributaries by smaller and less transparent dam developers.

For INGOs and other regional civil society organizations, the campaigns against the Xayaburi and other Laos dams have arguably failed. The dams are, after all, still being built. But this is not to say that these campaigns have not influenced the construction and design of these dams, or yielded concessions for affected communities. On the contrary, the Lao state does often attempt to deflect criticism by responding to it, although not to the degree that INGOs or other civil society entities might like.

Notes

1 Ethnicity in Laos has its own narratives, at the heart of which is the idea that ethnic diversity opposes a national identity. See Ireson and Ireson, 1991.
2 Exports from the dam to Thailand were routed through Udon Thani in Thailand's northeast, headquarters for the 7(13) US Air Force and a variety of CIA activities. The massive electricity consumption of this base, Thi (1999) argues, prompted US willingness to fund the dam.
3 The World Bank's retrospective on the experience was titled 'Doing a Dam Better' (Porter and Shivakumar, 2011).
4 Under the NT2, a website was even created. The latter provides superficial information on most of Laos's dam building projects, but special attention is given to the NT2. See www.poweringprogress.org.
5 40 per cent of shares in NT2 are held by Electricité de France (EDF); Norway's Statkraft holds 20 per cent of shares in THPC. The only other dam in Laos that might be considered to have a Western shareholder is the Houay Ho. 67.25 per cent of shares in the Houay Ho Hydropower Company are held by GLOW, a Thai subsidiary of EDF Suez.

References

ADB (Asian Development Bank) (1997) 'Hydropower project to increase Lao PDR's GDP by 7 percent', *ADB Review*, Nov.–Dec., Manila: Asian Development Bank.

ADB (Asian Development Bank) (2013) 'Theun-Hinboun Hydropower Project (Loan 1329-Lao[SF])', www.adb.org/documents/theun-hinboun-hydropower-project-loan-1329-lao-sf, accessed Oct. 2014.

ADB and World Bank (2006) *World Bank–ADB Joint Working Paper on Future Directions for Water Resources Management in the Mekong River Basin: Mekong Water Resources Assistance Strategy (MWRAS)*, Vientiane: World Bank.

ANN (Asia News Network) (2012) 'Laos seeks approval of Mekong countries for Xayaburi Dam', 4 May, www.asianewsnet.net/news-30219.html, accessed Oct. 2014.

Asian Scientist (2011) 'Xayaburi Dam decision delayed, pending further studies', 8 Dec., www.asianscientist.com/topnews/xayaburi-dam-northern-laos-delayed-mekong-river-commission-2011, accessed Oct. 2014.

Aubertin, C. (2001) 'Institutionalizing duality: Lowlands and uplands in the Lao PDR', *IIAS Newsletter*, 24, Feb., p. 11.

Baird, I. G., and Shoemaker, B. (2007) 'Unsettled experiences: Internal resettlement and international aid agencies in Laos', *Development and Change*, 38(5): 865–88.

Bangkok Post (2011) 'Xayaburi dam work begins on sly: Thai construction company, Laos ignore Mekong concerns', 17 April, https://groups.yahoo.com/neo/groups/coastfishclub/conversations/messages/1588, accessed Oct. 2014.

Bangkok Post (2012) 'Thai govt supports Xayaburi Dam', 6 Nov., www.bangkokpost.com/breakingnews/319838/thailand-backs-xayaburi-dam, accessed Oct. 2014.

Baran, E., Larinier, M., Ziv, G., and Marmulla, G. (2011) *Review of the Fish and Fisheries Aspects in the Feasibility Study and Environmental Impact Assessment of the Proposed Xayaburi Dam on the Mekong Mainstream*, Gland, Switzerland: World Wildlife Fund.

Barney, K., and Canby, K. (2011) *Baseline Study 2, Lao PDR: Overview of Forest Governance, Markets and Trade*, Forest Trends for FLEGT Asia Regional Program, July, www.forest-trends.org/documents/files/doc_2920.pdf, accessed Oct. 2014.

Bartlett, A. (2012) *Report for SDC: Trends in the Agriculture and Natural Resource Management Sectors of the Lao PDR*, Vientiane: Swiss Agency for Development and Cooperation.

BBC (British Broadcasting Corporation) (2012a) 'Laos' work on the Mekong River draws criticism', 4 July, www.bbc.co.uk/news/world-asia-18686438, accessed Oct. 2014.

BBC (British Broadcasting Corporation) (2012b) 'Laos approves Xayaburi "mega" dam on Mekong', 6 Nov., www.bbc.co.uk/news/world-asia-20203072, accessed Oct. 2014.

Bruce St John, R. (2006) 'The political economy of Laos: Poor state or poor policy?', *Asian Affairs*, 37(2): 175–91.

Costanza, R., Kubiszewski, I., Paquet, P., King, J., Halimi, S., Sanguanngoi, H., Bach, N. L., Frankel, R., Ganaseni, J., Intralawan, A., and Morell, D. (2011) *Planning Approaches for Water Development in the Lower Mekong Basin*, Chiang Rai: Portland State University and Mae Fa Luang University, July.

Defour, R. (1994) 'The Lao-American project and shifting cultivation', in D. Van Gansberghe and R. Pals (eds), *Shifting Cultivation Systems and Rural Development in the Lao PDR*, Report of the Na bong technical meeting, 14–16 July 1993, Vientiane: Nabong Agriculture College Project UNDP/DDSMS/LAO/92(017), pp. 229–34, http://lad.nafri.org.la/fulltext/1492-0.pdf#page=70, accessed Oct. 2014.

DW (Deutsche Welle) (2012) 'Xayaburi dam project proceeds as protest grows', 29 Aug., www.dw.de/xayaburi-dam-project-proceeds-as-protest-grows/a-16203873, accessed Oct. 2014.

Eco-Business (2012) 'Thai lawsuit threatens to derail Laos plans for Mekong river dam', 7 Aug., www.eco-business.com/news/thai-lawsuit-threatens-to-derail-laos-plans-for-mekong-river-dam, accessed Oct. 2014.

Evans, G. (2002) *A Short History of Laos: The Land In Between*, Crows Nest, NSW: Allen & Unwin.

FIVAS (Association for International Water and Forest Studies) (1996) *More Water, More Fish? A Report on Norwegian Involvement in the Theun-Hinboun Hydropower Project in Lao PDR*, Oslo: FIVAS.

Forsyth, T. (2003) *Critical Political Ecology: The Politics of Environmental Science*, London: Routledge.

Forsyth, T. (2005) 'Land use impacts on water resources: Science, social and political factors', in M. Anderson (ed.), *Encyclopedia of Hydrological Sciences*, Chichester: Wiley, pp. 2911–24.

Fullbrook, D. (2010) 'Food as Security', *Food Security*, 2(1): 5–20.

Goldman, M. (2001) 'Constructing an environmental state: Eco-governmentality and other transnational practices of a "Green" World Bank', *Social Problems*, 48(4): 499–523.

Hajer, M. (1995) *The Politics of Environmental Discourse*, Oxford: Clarendon Press.

Harvey, D. (1993) 'The nature of the environment: The dialectics of social and environmental change', in R. Miliband and L. Panitch (eds), *Real Problems, False Solutions*, London: Merlin Press, pp. 1–51.

Herbertson, K. (2011) *Sidestepping Science: Review of the Pöyry Report on the Xayaburi Dam*, Berkeley, CA: International Rivers, www.internationalrivers.org/files/attached-files/intl_rivers_analysis_of_poyry_xayaburi_report_nov_2011.pdf, accessed Oct. 2014.

Herbertson, K. (2013) 'Mekong countries at odds over mega-dams', 4 Feb., www.internationalrivers.org/resources/mekong-countries-at-odds-over-mega-dams-7824, accessed Oct. 2014.

Hirsch, P. (2002) 'Global norms, local compliance and the human rights–environment nexus: A case study of the Nam Theun II Dam in Laos', in L. Zarsky (ed.), *Human Rights and the Environment: Conflicts and Norms in a Globalizing World*, London: Routledge, pp. 147–71.

Hodgdon, B. D. (2008) 'Frontier country: The political culture of logging and development on the periphery in Laos', *Kyoto Journal*, 69: 58–65.

ICEM (International Center for Environmental Management) (2010) *Strategic Environmental Assessment of Hydropower on the Mekong Mainstream*, Vientiane: Mekong River Commission Secretariat.

ICEM (International Center for Environmental Management) (2011) 'Xayaburi HPP: Gains and losses for the LMB: Comments on the Xayaburi HPP Compliance Report: with a focus on the Mekong Delta', presentation delivered to the VUSTA Seminar, 23 Nov., www.icem.com.au/documents/envassessment/mrc_sea_hp/VUSTA%20poyry%20review.pdf, accessed Oct. 2014.

Ireson, C. J., and Ireson, W. R. (1991) 'Ethnicity and development in Laos', *Asian Survey*, 31(10): 920–37.

IUCN (International Union for the Conservation of Nature) (2013) 'After Xayaburi, it's time for some "hydro-diplomacy"', 5 Feb., www.iucn.org/news_homepage/all_news_by_region/news_from_asia/?11853/After-Xayaburi-its-time-for-some-hydro-diplomacy, accessed Oct. 2014.

Jusi, S. (2011) 'Challenges in developing hydropower in Lao PDR', *International Journal of Development*, 10(3): 251–67.

Kummu, M., and Varis, O. (2007) 'Sediment-related impacts due to upstream reservoir trapping, the Lower Mekong River', *Geomorphology*, 85(3–4): 275–93.

Kummu, M., Lu, X. X., Wang, J. J., and Varis, O. (2010) 'Basin-wide sediment trapping efficiency of emerging reservoirs along the Mekong', *Geomorphology*, 119: 181–97.

Lawrence, S. (2009) 'The Nam Theun 2 controversy and lessons for Laos', in F. Molle, T. Foran, and M. Käkönen (eds), *Contested Waterscapes in the Mekong Region: Hydropower, Livelihoods and Governance*, London: Earthscan, pp. 81–113.

Lao People's Democratic Republic (2006) *National Policy: Environmental and Social Sustainability of the Hyropower Sector in Lao PDR*, implemented by Science Technology and Environment Agency (STEA), Lao Environment and Social Project, Vientiane: Government of the Lao PDR.

Leap, J. (2012) 'In the Mekong, science – not guesswork – must prevail', Gland, Switzerland: WWF Global, wwf.panda.org/who_we_are/wwf_offices/laos/newsrom/?206911/In-the-Mekong-science-not-guesswork-must-prevail, accessed Oct. 2014.

Lestrelin, G. (2010) 'Land degradation in the Lao PDR: Discourses and policy', *Land Use Policy*, 27: 424–39.

Lestrelin, G., Vigiak, O., Pelletreau, A., Keohavong, B., and Valentin, C. (2012) 'Challenging established narratives on soil erosion and shifting cultivation in Laos', *Natural Resources Forum*, 36(2): 63–75.

Matthews, N. (2013) 'Drivers and enablers of hydropower development in the Lao PDR: The political ecology of Mekong riparians, investors and the environment', unpublished PhD thesis, Department of Geography, King's College London.

Mekong Committee (Committee for the Coordination of Investigations of the Lower Mekong Basin (1970) *Report on Indicative Basin Plan: A Proposed Framework for the Development of Water and Related Resources of the Lower Mekong Basin*, Bangkok: Mekong Committee.

Mirumachi, N., and Torriti, J. (2012) 'The use of public participation and economic appraisal for public involvement in large-scale hydropower projects: Case study of the Nam Theun 2 Hydropower Project', *Energy Policy*, 47: 125–32.

Molle, F., Foran, T., and Floch, P. (2009) 'Introduction: Changing waterscapes in the Mekong Region – historical background and context', in F. Molle, T. Foran, and M. Käkönen (eds), *Contested Waterscapes in the Mekong Region: Hydropower, Livelihoods and Governance*, London: Earthscan, pp. 1–19.

MPI (Ministry of Planning and Investment) (2011) *The Seventh Five-Year National Socio-economic Development Plan (2011–2015)*, Vientiane: MPI.

MRC (Mekong River Commission) Secretariat (2011) *Prior Consultation Project Review Report: Proposed Xayaburi Dam Project – Mekong River*, Vientiane: MRC Secretariat, www.mrcmekong.org/assets/Publications/Reports/PC-Proj-Review-Report-Xaiyaburi-24-3-11.pdf, accessed Oct. 2014.

MRC (Mekong River Commission) Secretariat (2013) *The BDP Story: The Story behind the Basin Development Plan*, Vientiane: MRC Secretariat, www.mrcmekong.org/assets/Publications/basin-reports/BDP-Story-2013-small.pdf, accessed Oct. 2014.

Nation (2012) 'Residents demand answers on Xayaburi', 2 May, www.nationmultimedia.com/national/Residents-demand-answers-on-Xayaburi-30181095.html, accessed Oct. 2014.

New York Times (2011) 'Decision looms for Laos dam, but impact is unclear', 17 April, www.nytimes.com/2011/04/18/world/asia/18mekong.html?module=Search&mab Reward=relbias%3Ar&_r=0, accessed Oct. 2014.

Pasaxon (2014) 'Causes of Kaluem District relocation', 12 March (in Laotian), www. pasaxon.org.la/Economic/12-3-14/Content9.html, accessed Oct. 2014.

Phnom Penh Post (2011a) 'US Senate pushes for Xayaburi funds freeze', 1 Dec., www. phnompenhpost.com/national/us-senate-pushes-xayaburi-funds-freeze, accessed Oct. 2014.

Phnom Penh Post (2011b) 'Laos goes "rogue" on dam', 24 June, www.phnompenhpost. com/national/laos-goes-rogue-dam, accessed Oct. 2014.

Phnom Penh Post (2012) 'Xayaburi concerns "unanswered"', 14 Sept., www. phnompenhpost.com/national/xayaburi-concerns-'unanswered', accessed Oct. 2014.

Porter, I. C., and Shivakumar, J. (2011) *Doing a Dam Better: The Lao People's Democratic Republic and the Story of Nam Theun 2*, Washington, DC: World Bank.

Reuters (2012a) 'Laos confirms has suspended controversial Xayaburi dam', 13 July, www.reuters.com/article/2012/07/13/laos-dam-idUSL3E8ID2AN20120713, accessed Oct. 2014.

Reuters (2012b) 'Laos holds groundbreaking ceremony for contentious Mekong dam', 7 Nov., www.reuters.com/article/2012/11/07/us-laos-dam-idUSBRE8A618I20121107, accessed Oct. 2014.

RFA (Radio Free Asia) (2012a) 'Xayaburi Dam redesign mulled: A new French study on the Mekong dam predicts no environmental impact, a senior Lao official says', 16 May, www.rfa.org/english/news/laos/xayaburi-05162012180613.html, accessed Oct. 2014.

RFA (Radio Free Asia) (2012b) 'Cambodia lodges dam protest with Laos: The Xayaburi hydropower project on the Mekong River stirs up controversy among Laos's neighbors', 1 May, www.rfa.org/english/news/laos/xayaburi-05012012190456.html, accessed Oct. 2014.

Rigg, J. (2005) *Living with Transition in Laos: Market Integration in Southeast Asia*, London: Routledge.

Roe, E. (1991) 'Development narratives, or making the best of blueprint development', *World Development*, 19(4): 287–300.

SEAWATER (South East Asia Water Business) (2012) 'Laos announces "postponement" of Xayaburi Dam but vows to continue construction and resettlement activities', 19 July, www.seawaterbiz.com/2012(07)/19/laos-announces-postponement-of-xayaburi-dam-but-vows-to-continue-construction-and-resettlement-activities, accessed Oct. 2014.

Shoemaker, B. (1998) *Trouble on the Theun-Hinboun: A Field Report on the Socio-Economic and Environmental Effects of the Nam Theun-Hinboun Hydropower*, Berkeley, CA: International Rivers.

Singh, S. (2009) 'World Bank-directed development? Negotiating participation in the Nam Theun 2 Hydropower Project in Laos', *Development and Change*, 40(3): 487–507.

Stuart-Fox, M. (2005) 'Politics and reform in the Lao People's Democratic Republic', Asia Research Centre Working Paper 126, ARS, University of Queensland.

Stuart-Fox, M. (2013) *Historical Dictionary of Laos*, 3rd edn, Lanham, MD: Scarecrow Press.

Thanh Nien News (2011) 'Vietnam hails Laos for suspending Xayaburi dam', 8 May, www. thanhniennews.com/index/pages/20110508124455.aspx, accessed Oct. 2014.

Thanh Nien News (2012) 'Diplomacy be damned: Work proceeds on the controversial Xayaburi dam as Thai contractor ignores agreement to suspend construction', 29

June, www.thanhniennews.com/index/pages/20120629-diplomacy-be-damned.aspx, accessed Oct. 2014.

The Diplomat (2012) 'Xayaburi Dam controversy flares: Construction on the Xayaburi Dam was supposed to be postponed under a deal last year. Has it been?', 20 April, http://thediplomat.com/2012(04)/xayaburi-dam-controversy-flares, accessed Oct. 2014.

The Economist (2012) 'Lies, dams and statistics', 26 July, www.economist.com/blogs/banyan/2012/07/mekong-river, accessed Oct. 2014.

Thi, T. D. (1999) *The Mekong River, and the Struggle for Indochina: Water, War and Peace*, Westport, CT: Greenwood Publishing Group.

Thoummavongsa, S. and Bounsou, X. (2013) 'Sustainable management of hydropower in Lao PDR', *GMSARN International Journal*, 7: 1–12.

THPC (Theun-Hinboun Power Company) (2014) 'History', www.thpclaos.com/index.php?option=com_content&view=article&id=39&Itemid=200&lang=en, accessed Feb. 2014.

Time (2011) 'Can damming the Mekong power a better life to Laos?', 12 Aug., http://content.time.com/time/world/article/0,8599,2088013,00.html, accessed Oct. 2014.

UPI (United Press International) (2012) 'Xayaburi Dam construction to continue?' 25 April, www.upi.com/Business_News/Energy-Resources/2012(04)/25/Xayaburi-Dam-construction-to-continue/UPI-47931335371706, accessed Oct. 2014.

Usher, A. D., and Ryder, G. (1997) 'Vattenfall abroad: Damming the Theun River', in A. D. Usher (ed.), *Dams as Aid: The Political Anatomy of Nordic Development Thinking*, London: Routledge, pp. 75–104.

Vaidyanathan, G. (2011) 'Dam controversy: Remaking the Mekong', *Nature*, 478: 305–7.

Varchol, D. (2012) *Mekong*, film at www.youtube.com/watch?v=ci_0L55_WEA&list=UUgMJ-RnHMc6ZvzAYDrfzFfw&feature=c4-overview, accessed Oct. 2014.

WCD (World Commission on Dams) (2000) *Dams and Development: A New Framework for Decision-Making*, Report of the World Commission on Dams, London: Earthscan.

WCED (World Commission on Environment and Development) (1987) *Our Common Future*, Oxford: Oxford University Press for the WCED.

Whitington, J. (2012) 'The institutional condition of contested hydropower: The Theu Hinboun-International Rivers collaboration', *Forum for Development Studies*, 39(2): 231–56.

Wong, S. M. T. (2010) 'Making the Mekong: Nature, region, postcoloniality', DPhil dissertation, Ohio State University, Columbus, OSU.

WWF (World Wildlife Fund) (2011) *Critical Review of the Pöyry Compliance Report about the Xayaburi Dam and the MRC Design Guidance: Fish and Fisheries Aspects*, Gland, Switzerland: WWF, http://assets.panda.org/downloads/review_of_fisheries_aspects_in_the_poyry_report.pdf, accessed Oct. 2014.

WWF (World Wildlife Fund) (2012) 'Laos in game of high stakes as it pushes ahead with Xayaburi dam construction', 24 July, http://wwf.panda.org/wwf_news/?205754/Laos-in-game-of-high-stakes-as-it-pushes-ahead-with-Xayaburi-dam-construction, accessed Oct. 2014.

Ziv, G., Baran, E., Nam, S., Rodríguez-Iturbe, I., and Levin, S. A. (2012) 'Trading-off fish biodiversity, food security and hydropower in the Mekong River Basin', *Proceedings of the National Academy of Sciences*, 109(15): 5609–14.

7 Whose risky business?

Public–private partnerships, build-operate-transfer and large hydropower dams in the Mekong Region

Carl Middleton, Nathanial Matthews and Naho Mirumachi

Introduction

The role of state and private sector in constructing large hydropower dams in the Mekong Region has shifted. Early projects since the 1960s in Thailand and Vietnam were principally conceived, built and operated by the relevant state agencies, typically with funding from the World Bank (WB) or the United States, and the Soviet Union respectively (Hirsch, 2010). Since the early 1990s, however, partial liberalization of the region's power sector (Middleton *et al.*, 2013), alongside regional economic integration and regional power trade steered by the Greater Mekong Subregion (GMS) programme, has increased the role of private-sector energy and construction companies and financiers (Middleton *et al.*, 2009). Whereas in Thailand and Vietnam, all of the largest hydropower dam projects have been developed already as state-led projects, in Laos and Cambodia, which are both presently undergoing extensive and rapid hydropower development, public–private partnerships (PPP) and build-operate-transfer (BOT) are the principal infrastructure investment vehicles. National policies and laws have facilitated this transition, promoted in particular by the WB, together with the International Finance Corporation, and the Asian Development Bank (ADB). The involvement of these actors and changes to policy reflect a global turn towards neoliberalism since the late 1970s in Northern countries, notably the United States, the United Kingdom, Germany and Australia, that led to a roll-back of the state and the emergence of the PPP and BOT model for infrastructure development there, closely tied up with privatization processes of state activities.

Thailand and Vietnam's power sectors have been partially liberalizing since the early 1990s and early 2000s respectively, allowing a greater role for independent power producers (IPPs) in domestic power markets; in Thailand and Vietnam, 40 and 47 per cent of the total electricity capacity is produced by IPPs respectively (Nguyen, 2012; EPPO, 2013). These policies have also opened up opportunities for power imports by IPPs from neighbouring countries. Thailand and Vietnam have increasingly looked to their neighbouring countries for power imports due to a range of considerations, including rising electricity demand, exhaustion of the most suitable domestic hydropower sites and growing domestic controversy around the social and environmental impacts of hydropower projects; Thailand plans to import approximately 12,000 MW of coal-fired electricity and hydroelectricity from Myanmar and Laos by 2030 (EPPO, 2012), while Vietnam plans to import 2200 MW by 2020 and 7000 MW by 2030 from Laos, Cambodia and China (Socialist Republic of Vietnam, 2011). Meanwhile, private energy and construction companies from Thailand and Vietnam, alongside state-owned and private companies from China, Malaysia, and Russia among others, have sought to develop hydropower projects in Laos and Cambodia for both these country's domestic markets and for power export.

This chapter adopts a political economy approach to examine how private-sector investment, and more specifically PPPs, IPPs and the BOT model, have shaped the hydropower development process in the Mekong Basin and the approach of government, investors and developers. We highlight the role of national and regional institutions and a range of actors including: the state (politicians and bureaucrats); civil society (local and transnational); private sector (financiers, and the construction and energy industry); international financial institutions; and bilateral aid agencies. Sayer (1985) argues that actors derive power from and act according to the structures in which they are situated. This chapter will evaluate which actors have derived power from the BOT/ PPP model. The chapter examines in particular the emergence and growing role of IPPs and private sector financing in Laos and Cambodia. The chapter shows how the divergent political economies of the two countries have resulted in different models of BOT/ PPP in large-scale hydropower development. Moreover, analysis of the political economy provides an insight into the ways in which state and non-state actors engage in water governance.

The chapter argues that the actual distribution of project risk and benefit does not reflect the original rationale of BOT/ PPP (see also Wyatt, 2004), and in particular leaves the state, local communities, the environment and electricity consumers with a disproportionate share of risk, to the benefit of the private sector developers and financiers (see also Greacen and Greacen, 2004). It is also pointed out that the 'classic' actors such as the WB and the ADB still stand to gain from the projects and are not entirely irrelevant despite the influx of new private sector actors (see also Middleton *et al.*, 2009). Yet, the governments of Cambodia and Laos also struggle to negotiate deals as they compete to attract international and regional capital. Overall, it is argued that the interests of IPPs are a significant factor driving hydropower development in the Mekong Basin, shaping energy planning in Thailand and Vietnam, and relevant law and policy on hydropower development in Cambodia and Laos.

In the following section, a brief review of the principles and origins of the BOT and PPP model is provided. This is followed by an analysis of the political economy of Laos and Cambodia power sectors, with a focus on hydropower development. In these sections, the origin of the BOT/ PPP model for each country is discussed, together with how each country's political economy has shaped the legal form and institutional basis of BOT/PPP models that have emerged. Each section also considers the performance of the BOT/PPP model to date, in terms of economic, social and environmental performance, transparency and accountability to civil society and locally affected people, and distribution of technical, financial and political risks. The final section provides a brief synthesis and conclusion.

The rise of the BOT model

The BOT model is well known around the world. The first BOT project was the Suez Canal in 1868. The model became increasingly popular with the deregulation and liberalization of Northern market economies from the 1970s to the 1990s. BOT model projects were promoted by multilateral development agencies as a way to encourage private sector financing of large-scale infrastructure projects such as the Channel Tunnel between Britain and France and Kansai International Airport in Japan.

The BOT model concept involves the private sector handling all the financing, design, construction and operation of what would previously typically have been a public infrastructure project for a concessionary period of usually 20–50 years (Levy, 1996). During the contracted period the private operator is allowed to run the infrastructure at a rate of return high enough to service debts within a maturity period, the time in which the debt of the infrastructure remains outstanding, and afterwards to generate a profit of approximately 15 per cent or more (Levy, 1996). Once the concessionary period is finished the government gains ownership of the infrastructure at no cost.

In the Mekong Basin's hydropower development, the BOT model has been deployed in a similar way. For example, in the case of Laos's BOT projects, the government of Laos (GoL) – usually with donor support – has retained a percentage of control in projects throughout the concessionary period by becoming part of the consortium of investors. This PPP arrangement has allowed investors to exert a degree of control over the project depending on what stake they take. In many dam projects, the GoL has also involved its state-owned enterprises (SOEs) or government departments in the environmental and social mitigation measures. In the Theun-Hinboun project, for example, the GoL took a 60 per cent majority stake in the project, while a military-owned company was involved in logging the reservoir.

BOT structures tend to be complex. Within the BOT contract there are dozens of fees, guarantees, loans and contracts needed between each actor. Figure 7.1 shows a typical simplified BOT structure for a hydropower plant.

Distributing risk

Because the nature of BOT projects restricts investors from removing their equity when they please, they generally assemble a large consortium of investors, including state-backed banks, and attempt to spread their risk exposure. The

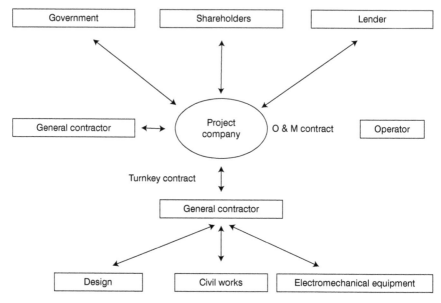

Figure 7.1 BOT Structure for Hydropower Plant

Source: Forouzbakhsh *et al.*, 2007.

commercial viability of the project needs to be carefully assessed by the project financiers as they provide the bulk of investment and have limited recourse to recoup losses should the project fail. While BOT projects should encourage careful investments because of the substantial risks involved, opportunities for investors to profit through construction or service supply may incentivize investments with weak returns. For example, Morris (1997) details a BOT case in Australia where a private company published misleading revenue estimates from the Melbourne Citylink Project to attract investment. With investment secured, decision-making on the project could move forward. As the project proceeded, numerous companies involved in the construction and service supply of the project were able to profit from its development despite the questionable longer term returns of the project itself. Thus, uncertainty lies in how revenues can be identified and assessed; revenue estimates can be redrawn based on actors' incentives for a quick profit. The intention of this chapter is to point out the nature of risks and uncertainty in BOT/PPPs in the Mekong context.

Furthermore, risk may be poorly understood or underestimated by investors. Handley (1997) claims that governments, particularly from developing countries, have been encouraged to absorb risks and provide private sector guarantees, such as supplying security, water flows, fuel or electricity. These risks are rarely adequately costed. The complexity of projects results in the possibility of a range of environmental, socio-economic, financial and political risks being undervalued, overlooked or misidentified. In developing countries, risks may be further exacerbated by poor-quality data or baseline studies. For example, hydropower projects that rely on past hydrological data to predict future flow rates and thus project viability may need to extrapolate conclusions on project viability from limited baseline data, which may also affect environmental and social impact mitigation measures. As will be discussed below in case studies from Laos and Cambodia, the risks associated with BOT hydropower investments are particularly high in the Mekong because the construction of hydropower dams causes immediate and long-term environmental and social impacts well beyond the borders of the country developing hydropower.

BOT and IPP in Laos

The emergence of the BOT model in Laos and the influx of private-sector investment have had significant implications on how hydropower is developed and planned across the country. Promoted in Laos in the early 1990s by the World Bank, the ADB, the United States and others, the purpose of introducing private sector investment and the BOT model to Laos was twofold. First, private-sector investment and BOTs provided an opportunity for the GoL to develop complex and costly projects that the state did not have the capacity to finance or construct (Wyatt, 2004); and second, IPPs and BOT projects were argued to provide a suitable vehicle to open up Laos markets to international and regional investment and development. The opening up of the Laos market was part of a broader neoliberal agenda to bring the country out of poverty through private-sector investment, market expansion and rapid infrastructure development. Dam construction and electrification led by private sector investment have been central to this vision of development and modernization for Laos and for the region in general since the Mekong Committee first proposed projects in the 1960s (Nguyen, 1999).

The first hydropower project in Laos to trial this new private sector-led model was the 45 MW Xeset Dam completed in 1991. The ADB supported the funding of the dam as a key step in opening up of the Laos economy (Cruz-del Rosario, 2011). This dam and the Nam Ngum Dam, completed in 1971, were the first to export surplus energy to Thailand, laying the groundwork for future exchanges with Thailand and the region.

As a result of the completion of the Xeset Dam, the Ministry of Industry and Handicrafts (later renamed the Ministry of Energy and Mines) began to receive dozens of Memorandum of Understanding (MoU) requests from regional and international private hydropower developers keen to profit from Laos's newly opened hydropower industry (see Table 7.1).[1]

IPPs were an important force in shaping the Laos economy. As one consultant put it, 'The IPP model was about the only economic card the GoL had to play' (consultant, pers. comm., 2011). The opening up of the Laos economy to the private sector represented more than hydropower development. The neoliberal policies exposed the underdeveloped economy of Laos to a host of global and regional forces. Investors were keen to profit from Laos's relatively untouched

Table 7.1 Hydropower projects and MoUs in Laos as of November 2003

Project	Capacity	Project company / lead Sponsor	Type of agreement	Power purchase agreement
Theun Hinboun	210	Theun Hinboun Power Company	Concession Agreement	Yes
Houay Ho	150	Houay Ho Power Company (Daewoo – South Korea)	Concession Agreement	Yes
Nam Theun 2	720	Nam Theun Electricity Company	Concession Agreement	Yes
Nam Ngum 2	980	Shlapak (US)	Concession Agreement	No
Nam Ngum 3	440	GMS Power (Thailand)	Project Development November 1997	No
Sepian/Senamnoy	390	Dong Ah (South Korea)	Concession Agreement	No
Xekaman 1	468	ALP Management (HECEC – Australia)	Concession Agreement	No
Nam Theun 3	237	Heard Energy (US)	Project Development Agreement 1994	No
Nam Mo	105	Mahawongse	Project Development Agreement November 1999	No
Nam Tha 1	263	SPS	Memorandum of Understanding October 1995	No
Nam Theun 1	540	SUSCO (Thailand)	Memorandum of Understanding March 1994	No
Nam Lik	100	Hainan SIT (China)	Memorandum of Understanding February 1994	No
Nam Ngum 5	90	Melkyma	Memorandum of Understanding September 1996	No

Project	MW	Developer	Agreement	Operational
Nam Ou	600	Pacific Rim	Memorandum of Understanding November 1994	No
Xe Katam	100	Hydro Power	Memorandum of Understanding October 1994	No
Nam Khan 2	126	Hydro Quebec (Canada)	Memorandum of Understanding 1994	No
Nam Suang 2	190	VKS	Memorandum of Understanding March 1995	No
Nam Nhiep 2+3	565	VKS	Memorandum of Understanding March 1995	No
Phapheng (Thakho)	30	True Assess Ltd.	Memorandum of Understanding	No
Nam Bak (Cha) 2B	120	Nisho Iwai (Japan)	Memorandum of Understanding	No
Nam Bak (Cha) 1+2	185	HECEC (Australia)	Memorandum of Understanding	No
Xe Kong 4	528	Modular	Memorandum of Understanding	No
Nam Ngiep 1	440	Shlapak (US)	Memorandum of Understanding	No
Nam Mang 3	50	Ch Kanchang	Memorandum of Understanding	No

Source: interviews and Wyatt's unpublished PhD thesis on BOOT in Laos and Vietnam (2004).

market and natural resources, not only water resources but also mining and forestry. One of the key drivers of this investment during the early 1990s was the economic boom in Asia. In the years preceding the Asian Financial Crisis, Asian economies were developing rapidly and many firms were seeking new markets and high-return, high-risk investments for large amounts of capital (Studwell, 2007). A WB report states that from 1990 to 1997 global private annual investments in developing country infrastructure rose from $19 billion to approximately $120 billion (Izaguirre and Rao, 2000).

In the case of Laos, the surge of investment that accompanied the opening up of the country in the 1990s presented a number of challenges to the GoL. Limited laws and policies were in place to protect against hydropower's environmental and social impacts and the GoL had inadequate experience dealing with private-sector investors and the array of risks associated with these large-scale developments. During this period, the pressure on Laos to rapidly develop its infrastructure and open its economy to private sector investment resulted in a number of exploitative hydropower developers entering into agreements with the GoL (GoL official, pers. comm., 2011).

Houay Ho Dam

The Houay Ho (150 MW) Dam provides an early case of poor practice in how private-sector investment and the BOT model proceeded in Laos. Houay Ho's environmental impact assessment was only completed two years after construction began. According to consultants who examined the dam for the ADB post-construction, the dam was built on overly optimistic hydrological flows that resulted in its power production being below anticipated performance (consultant, pers. comm., 2012), contributing to the fact that its electricity production rights were sold four times in the 11 years following its construction. As discussed in this section, throughout the life of the dam to 2013, developers have consistently attempted to avoid taking responsibility for its impacts. Furthermore, in the rush to the build the dam, the GoL appears to have significantly misjudged the financial risks of the project, by exposing itself to heavy interest payments with little visible return.

Houay Ho is located in the south of the country 160 km east of Pakse and 30 km northwest of Attapeu. It was completed as a BOT project in 1998 with Daewoo Engineering and Construction Company owning 60 per cent, the state-owned Electricité du Laos (EdL) owning 20 per cent and Loxely PLC (a Thai development firm) owning 20 per cent. In 1993, the Houay Ho Power Company (HHPC) was created to develop the project. Thailand's electricity utility, EGAT, was to buy 95 per cent of the electricity which is exported to Thailand. In 1998, shortly after the dam was completed, Daewoo and Loxely PLC sold their shares. Tractebel Electricity and Gas International, a Belgian firm, and MCL, a Thai firm, joined to purchase 80 per cent of the Daewoo and Loxely shares for the dam (including debts), with the GoL retaining the remaining equity (Houay Ho, 2011).

The project was sold again in 2002 with GDF Suez buying 60 per cent of Tractebel and MCL's share. In 2009, it was sold a third time when a subsidiary of GDF, Glow Energy purchased 67 per cent of GDF's shares (Houay Ho, 2011). The current ownership structure sees Glow Energy with 67.25 per cent of shares, EdL with 20 per cent, and Hemaraj Land and Development with 12.75 per cent (Houay Ho, 2011). Glow Energy is a Thai IPP and Thailand's third largest energy producer, supplying 7–8 per cent of the country's electricity (Houay Ho, 2011). The continual reselling of the dam speaks to the financial climate, poor hydropower potential of the dam and the mounting criticism of the developers due to the weak social and environmental impact mitigation measures.

Since its inception, the Houay Ho developers and the GoL have managed to circumvent many of the environmental and social protection laws and policies within the country. Houay Ho's EIA was completed two years after construction commenced (Khamin, 2008). The project resulted in the resettlement of 4,000 Heuny and Jrou ethnic minority people with inadequate compensation (Khamin, 2008). For example, only 20 per cent of the land originally allocated for compensation turned out to be available, as the remainder was already in use by other villages (Khamin, 2008). This resulted in serious food security impacts for the resettled people (Khamin, 2008). An observer from the former Ministry of Industry and Handicrafts, GoL, was quoted as saying that '[i]t had a bad smell. We never got to see any studies for the project. I do not think any were done' (Khamin, 2008).

Tractebel, the company that owned the dam at the time the EIA was completed, denied responsibility for the earlier failures of the project (Khamin, 2008). INGO pressure led to a complaint against Tractebel in Belgium accusing them of disregarding the Organization for Economic Cooperation and Development's (OCED) Guidelines for Multinational Enterprises. As a result, the company undertook some limited resettlement improvements, including the building of a school and some wells (International Rivers, 2008).

In 2001, in response to concerns raised by INGOs, the ADB sent Electrowatt-PA Consulting to the Houay Ho site to complete an assessment of Tractebel's resettlement improvements. The ADB did not appear to be originally involved in the project, but was advising the GoL on hydropower investments across the country. The report from Electrowatt-PA Consulting stated that a significant portion of the improvements had not been implemented (ADB, 2003). The report further highlighted a number of problems with the resettlement, including poor water quality and quantity, insufficient land for livestock grazing, illegal logging in the surrounding area and poorly equipped schools and clinics (ADB, 2003).

The dam illustrated severe failings in terms of the GoL's capacity to judge risks, and a neglect of Laos's environmental laws and policies, leading to significant environmental and social damage. Delang and Toro (2011) found that the GoL's weak capacity and poor legal representations during negotiations resulted in developers taking advantage of the political and economic environment in the country. This weak capacity within the approval process did not only benefit developers. Powerful actors in the GoL undoubtedly also benefited from the dam's weak social and environmental assessment, which has since been the object of reform (see Khamin, 2008).

The project did not produce any royalties for EdL until 2010, although the government was required to make annual interest payments of US$1.8 million on its $10 million dollar equity from 2000 (Delang and Toro, 2011). Due to the weak concession agreement, Daewoo, the original project developer, made a single payment of $230,000 for social and environmental impacts and left the GoL to deal with any subsequent issues (Delang and Toro, 2011). As of 2012, the project has ongoing difficulties meeting its electricity export agreements with EGAT, due to the overly positive assessment of water availability prior to construction, and resettlement disputes are ongoing (consultant, pers. comm., 2012).

Nam Theun 2

Another significant example of the PPP/BOT model and its implications for Laos's hydropower development is the Nam Theun 2 (NT2) Hydroelectric Project. The NT2 represents an important marker in the political economy of hydropower development in Laos because of its broad-reaching implications for existing modes of hydraulic investment. In many ways, the NT2 was the first test case for the assessment and regulation of financial, socio-economic and environmental considerations of large-scale hydropower projects. The NT2 project was set up with the Nam Theun 2 Power Company Limited (NTPC) as the project developer responsible for 25 years of management before the operation is transferred to GoL. A key to understanding the political economy of hydropower development through the example of the NT2 is the purpose of energy production: sale of electricity to Thailand. The project has been designed as one that could generate 1070 MW, making it the largest hydropower project in operation in Laos at the time of writing; 995 MW of this power (95 per cent) is exported to Thailand (NTPC, n.d.; De *et al.*, 2010).[2] The export model of hydropower power development ties the rationale, costs and benefits of the project to a transborder structure of energy supply and demand. The project is thus symbolic of the interest in utilizing the water resources for energy, with its sheer scale and significance to the Laos and Mekong economy.

NT2 involves a large range of private and state actors, in addition to international financial institutions and export credit agencies. The project, expected to raise US$1.9 billion as revenue for GoL, is supported by the investments of 16 commercial banks, consisting of nine international banks and seven Thai banks (NTPC, 2014a). The current project shareholders, EDF, Lao Holding State Enterprise (LHSE) and ECGO with 40, 35 and 25 per cent each in the shares of NTPC (NTPC, 2014b), represent a mix of global and regional energy sector characteristics. These three shareholders have been involved in the consortium that eventually received approval for NT2 since the beginning, but it is also worthwhile to note that a fourth shareholder of that consortium divested its shares to the remaining three shareholders shortly after project completion; the Thai construction company Ital-Thai Development initially held 15 per cent of NT2's shares, and was also awarded the role of main civil work contractor in the project, thus recouping its profit against risk at this early stage of the project cycle. Meanwhile, EDF – which was also the project's head contractor – represents an international brand with wide-ranging expertise, much needed for a large-scale project such as NT2. EGCO is a Thai IPP, representing a new set of actors emerging from the privatization efforts of the Thai energy sector. LHSE is a state-owned company that undertakes much of the initiatives of GoL to develop large hydropower projects for power exports. In order to facilitate the export of NT2's electricity, the power purchase agreement was signed between EGAT and EdL, bringing into the mix actors for energy generation and transmission.

From its inception, the NT2 has drawn on the resources and expertise of multiple actors, and the consortium of project developers has changed over time. The project was first given serious consideration in the mid-1980s when the Australian based Snowy Mountains Engineering Corporation (SMEC) began initial surveys of the project site, and received a commission for a feasibility study in 1991 from the GoL and two international organizations: the WB and the United Nations Development Programme (UNDP) (NTPC, 2005). The prospect of utilizing the mainstream and tributaries of the Mekong River Basin, including the Nam Theun River, has been one defined through macroeconomic development aspirations and initiatives (Bakker, 1999; Molle *et al.*, 2009). As pointed out by Sneddon and Fox (2012), the role of external agencies such as United Nations organizations, donor agencies and international financial institutions is not insignificant in the grand plans for development of the Mekong River Basin. While the interests and geopolitical drivers that engaged these actors are beyond the scope of this chapter, these organizations have mobilized significant financial resources and expertise. In the case of the NT2, the World Bank had been involved in investigating options for development in its tributary basin (Porter and Shivakumar, 2011) and the UNDP has been instrumental in the institutional development of water resources management (Mirumachi, 2012). In addition, as a project that was to export electricity across the border, Thai investment and infrastructure companies such as Ital-Thai, Merrill Lynch Phatra Thanakit Securities and Jasmine International initially registered their interest, forming a consortium to undertake the project with Transfield Holdings Ltd., an Australian construction and consultancy firm, and EDF (Hirsch, 2002).

A project development agreement (PDA) was signed in 1993 between this consortium and the GoL but the period until the realization of the concession agreement (CA) in 2002 reflects uncertainties and risks related to the scale of the project and to the capacity of the host government, as well as opposition to the project by a range of NGOs (Lawrence, 2009). Financing the project entirely with commercial banks proved to be challenging as these organizations perceived the length of lending of 30 years and sovereign risk as bottlenecks

(White, 2008; Porter and Shivakumar, 2011). To make the project financing robust, a series of arrangements were made to mitigate risk and deter the GoL from defaulting. A main component to such arrangements was the involvement of the World Bank. Largely at the request of the commercial banks, the World Bank became part of a financial structure set up to buffer against any financial loss from failures or non-compliance to the BOT agreement (Guttal, 2000). In addition, safeguards against political risk were also sought from the Multilateral Investment Guarantee Agency (MIGA) and export credit agencies (Cruz-del Rosario, 2011). The complexity and time-consuming nature of these arrangements is reflected in the change to the consortium of developers. Transfield Holdings Ltd who had led the consortium withdrew, resulting in EDF and EGCO playing a larger role (Porter and Shivakumar, 2011).

It is important to note the heterogeneity between actors in the ways they dealt with these risks and uncertainties. On one hand, the large-scale project did not fit the risk thresholds of commercial banks and delays to the CA are high opportunity costs to developers. On the other hand, the WB was concerned about whether the environmental and socio-economic issues would meet their safeguard policies. The Asian financial crisis in 1997 underscores this point on diversity. The financial crisis had a severe impact on many investment projects across the region, and not just water resources development projects in the Mekong, making the region generally unfavourable to high-risk projects such as the NT2 regardless of the actor. Indeed, the NT2 was put on hold at this time. However, the significance of this pause in the procedure for green-lighting the project has different meaning for various actors. For the Thai government, new electricity supply was not an urgent issue any more. In contrast, companies with a broad portfolio like EDF diversify their risks as a matter of business practice. Interestingly, while EDF maintained its interest in the project even after the Asian financial crisis, it decided to retract its participation temporarily in 2003, raising speculation over low confidence in the project (Lawrence, 2009). The Asian financial crisis came at a time when a major initiative to investigate better practice of dam project planning and implementation was set up. The World Commission on Dams (WCD) placed greater expectations on the World Bank for environmental and social performance, beyond its own emphasis on financial safeguards (Porter and Shivakumar, 2011).

The implications of the practice established through the NT2 are multi-faceted. First, the practice of environmental and socio-economic impact assessment is now regarded by the GoL as an exemplar for large-scale investment in Laos. The involvement of the WB was not a preferred approach initially because of the time and resources required (Hirsch, 2002; Porter and Shivakumar, 2011). However, the NT2 is presented by the GoL as a successful project that meets international standards. This is not to say that criticisms on the impact assessment and project evaluation do not exist. Developing large-scale dams is a divisive issue, with debates on the socio-economic and environmental implications, and NT2 was no exception. It has been argued that the environmental impact assessment was not comprehensive, poorly conducted and badly communicated (Goldman,

2005; Lamberts, 2008; Lawrence, 2009). Critiques have been made on the way crucial issues such as public participation and cost-benefit analysis were handled by influential actors such as the WB and ADB (Greacen and Sukkamnoed, 2005; Singh, 2009; Mirumachi and Torriti, 2012). Moreover, when considering the lengthy and complex process to finance and safeguard the NT2, it is doubtful if it is a model that can be easily replicated.

Second, the project has paved the way for a mode of hydropower development in which government agencies and state-owned companies have a large stake. The close links of these actors help establish narratives such as Laos as the 'battery of southeast Asia', claiming large hydropower dams as a viable source of economic development. The recent controversy over the Xayaburi Dam in Laos (Matthews, 2012), the first dam on the Mekong River's mainstream, reflects the dominance of this narrative. Third, and relatedly, hydropower for export ties the political economy of the Thai energy sector very closely to Laos energy development (Middleton *et al.*, 2013). While it can be argued that the involvement of Thai IPPs may bring about diversified options for development for Laos compared to the cumbersome, lengthy process of state-to-state development projects, the fundamental structure according to which the Laos energy sector is dependent on Thai energy market trends is not unimportant. Moreover, when details of the power exchange agreements are examined in detail, it becomes clear that this mode of hydropower development is not only about the energy security of Thailand or Laos, or for the Mekong Region. Hydropower development commodifies water resources as an extractive natural resource to benefit those with a stake in state-led development initiatives (see also Middleton *et al.*, 2013). As Hirsch (2002, 157) put it:

> it [NT2] should more readily be seen as an enclave resource extraction initiative, similar to a mining operation, whose principal raison d'être is to generate foreign exchange in the form of dividends on state-held shares and from resource royalties.

Matthews (2012) attests to this claim by highlighting the short-term nature of profit prospects by a select group of influential actors. Despite the variety in the ways risks and uncertainties are perceived by the investors, developers and the external agencies, these relate to project-specific planning and implementation rather than broader societal risks and normative uncertainties of development choices.

Policy and practice of IPP/BOT in Laos

These two examples point to several important insights. The GoL was ill-prepared for the transition to the involvement of IPPs and rapid BOT project expansion. With no political risk guarantees and poor risk-mitigation measures, many more reputable hydropower developers avoided Laos because the potential costs and risks outweighed the benefits (consultant, pers. comm., 2012). As an industry insider from the 1990s stated:

The lack of capacity in the Government meant that the objectives were not defined in any quantitative sense and they did not have the people and institutions to control a program of such size and complexity. There was over-reliance on developers for the financial modelling to project GoL benefits and contracts needed to mobilize money from the debt markets. They were lambs to the slaughter in the early years.

(Industry insider, pers. comm., 2012)

Huntington (1965) has pointed out that strong political institutions are necessary for the political institutionalization needed to regulate rapid modernization. Strong political institutions have the necessary scope, adaptability, complexity, endurance and coherence to regulate rapid development (Huntington, 1965). Many of these principles were absent in Laos because it had only recently emerged from its independence from the French and then witnessed its revolution and transition into its current communist one-party system.

To this end, with the onset of the Asian financial crisis in 1997, many projects outlined in Table 7.1 were placed on hold or scrapped. This period gave the GoL some breathing room to develop its capacity. Recognizing the lack of capacity within the GoL to regulate private-sector investment, the World Bank and the ADB worked closely with the government to put in place a number of laws and policies to regulate hydropower development and increase transparency. These included the Environmental Protection Law (1999) that requires all development projects that have the potential to affect the environment to prepare an environmental impact assessment, and Environmental and Social Safeguard Policies (2002) designed to minimize impacts. In many ways, these new legislative pieces are informative and innovative, not just for Laos, but for the region more generally. The new laws and policies implemented by the government, however, only had a minimal impact on regulating private-sector investment and the BOT model due to the business norms that accompany these investments. Hydropower development agreements between the private sector and the government have been subject to intellectual property rights and a host of other legal rules that protect the competitive interests of the private sector. These private-sector rules have restricted government influence on developments, decreased transparency and further politicized the decision-making environment. The influence of these private sector norms on transparency is evidenced in a WikiLeaks cable from 2008:

A number of NGOs have previously complained about the Government's [of Laos] refusal to abide by its own policy. When questioned about the policy, the Deputy Chief of Social and Environmental Management at the Ministry of Energy and Mines, appeared unaware of the national policy promoting hydropower transparency, and claimed that EIAs were private company documents.

(WikiLeaks, 2008)

As the example of the NT2 showed, apart from risks, the influx of private-sector investment and the BOT model also shaped the scale of hydropower development. Bakker (1999, 225) identified that BOT, private-sector-led projects have had 'a tendency to favour large-scale, capital intensive projects over smaller-scale initiatives' leading to 'a different prioritization of hydro projects than that which may have been determined by the Mekong River Commission or Government planning agencies'. Evidence of this can be seen in the number of MoUs signed by the GoL in Table 7.1. Only five out of 23 MoUs were for hydropower projects under 100 MW. Suhardiman *et al.* (2011) has also argued that the priorities of the private sector have been privileged over more broadly defined river basin planning objectives in Laos. It should also be pointed out that larger projects come with increased complexity and diverse risk, including engineering, financing and social and environmental impacts, not to mention uncertainty over timely completion. They also create larger opportunities for corruption through complex contract processes.

BOT and IPP in Cambodia

Cambodia's power generation is largely privately owned, reflecting the country's recent history and the evolution of power-sector policy since the early 1990s. Cambodia's electricity infrastructure, which before the 1970s included only one major transmission line, was seriously damaged in the civil war (Poch and Tuy, 2012). Subsequent political instability and underinvestment has left present-day Cambodia with an electricity system that is undependable and costly, and the Cambodian Government has prioritized the sector for investment and expansion. Cambodia's electrification rate is among the lowest in Southeast Asia and access to electricity is concentrated in urban areas (Chea, 2009). Electricity prices are high and volatile due to a heavy dependence on generators fuelled by expensive imported diesel and a high proportion of electricity imports from neighbouring countries (Poch and Tuy, 2012; see Chapter 8).

Since the mid-1990s, Cambodia has sought to rebuild its economy and electricity infrastructure. Cambodia does not at present have a nationwide grid; as of 2009, there were three interconnected power systems (Phnom Penh, Northwestern Grid and the Southern Grid) and two systems that import electricity from Thailand and Vietnam. The Ministry of Industry Mines and Energy (MIME) aims for the completion of a national grid by 2018 (REEEP Policy Database, 2012). Cambodia has power cooperation agreements with its neighbouring countries (Laos, Vietnam and Thailand) that facilitate Electricité du Cambodge (EDC), the state-owned utility, in importing power (EAC, 2013). The Electricity Authority of Cambodia (EAC), Cambodia's electricity regulator, has also authorized some private companies to purchase electricity directly from Thailand to sell to areas in Cambodia near the border (EAC, 2013). The MIME, EDC and EAC are the three main state agencies responsible for the electricity sector in Cambodia (see Chapter 8).

While national electricity consumption is comparatively low, demand outstrips supply and according to the government annual electricity demand growth rate is approximately 19 per cent (Jona, 2011). Domestic generation had increased 40 per cent over 2011 (EAC, 2013). By fuel type, 36 per cent is generated by hydropower, 60 per cent by diesel or heavy fuel oil, 3 per cent by coal and 1 per cent by wood or biomass (EAC, 2013). The government estimates that by 2020 national power demand will be at least 2770 MW, increasing from approximately 1060 MW in 2012 (CDC *et al.*, 2014).

All large hydropower projects in Cambodia have been developed as BOT projects by IPPs, with China a particularly heavy investor to date. In contrast to Laos, the government does not take a shareholding in the projects. EAC has issued a growing number of licences, from 21 in 2002 to 312 in 2012. For electricity generation licensed by the EAC, 92 per cent of electricity is generated by IPPs, 3 per cent by consolidated license holders[3] and 5 per cent by EDC (EAC, 2013). EDC is the only public-sector actor issued a licence, with all other licences issued to private sector actors engaged in electricity generation, transmission and distribution (EAC, 2013). EDC is responsible for electricity generation, transmission and distribution in 12 provincial capitals, Phnom Penh and neighbouring Kandal Province (CDC *et al.*, 2014); due to its limited financial resources even in these areas, generation is largely sourced from IPPs and power imports (CDRI and ANZ Royal Bank, 2010). Other areas of Cambodia are served by IPPs, various rural electricity enterprises (REEs), and companies licensed to import power from neighbouring countries (Poch and Tuy, 2012).

The key role played by the private sector reflects Cambodia's recent historical circumstances, and has also been shaped by policy advice from the World Bank. As reflected in Cambodia's Electricity Law (2001), the future development of Cambodia's electricity generation capacity and transmission and distribution infrastructure is to be led by private sector investment in the form of BOT, or by some large coal-fired power stations as build-operate-own (BOO) (Poch and Tuy, 2012). Similar to the Laos case, the World Bank has played a particularly influential role in formulating Cambodia's power sector policy, and is a keen supporter of the BOT model and private-sector-led expansion (e.g. World Bank, 2002, 2005, 2006), as has been the ADB (e.g. MIME, 2003, 90).

There are divergent visions for Cambodia's electricity sector. Ryder (2009, 51) argues that Cambodia's current policy for the electricity sector

> focuses on high-voltage transmission connections with neighbouring countries, large-scale hydro dams, and coal-fired plants (of unspecified technology). Excluded are entities other than big-scale power producers that want to generate power locally or install building-scale technologies that allow consumers to supply their own needs or reduce their demand for grid based electricity service during the day.

Ryder (2009) concludes that the Cambodian government considers REEs as an interim solution until the grid – powered by large IPPs – can be extended, and that REEs operate in a difficult business environment with little access to affordable capital, a lack of certainty in the long term for permission to operate, and with a lack of clear rules. Poch and Tuy (2012, 169) also suggest risk is quite high for investors, but conclude that 'only large-scale investments appear to be viable'.

Plans for hydropower in Cambodia

Large hydropower dam construction has been placed at the centre of Cambodia's power development plans, together with the construction of a high-voltage transmission network linking urban centres. Cambodia's hydropower potential is estimated to be 10,000 MW, of which 50 per cent is located on the Mekong River's mainstream, 40 per cent on the Mekong River's tributaries, and the remaining 10 per cent in the southwestern coastal area (Jona, 2011).

Before the arrival of Chinese hydropower developers, the Cambodian government had struggled to attract investment for large hydropower projects (Middleton, 2008). Despite their presence since the early 1990s, Western bilateral donors, the ADB and the World Bank, and private-sector Western hydropower developers have been reluctant to support large hydropower in Cambodia, including due to the environmental and social impacts and the country's weak governance. The recent growing closer of economic and political ties between China and Cambodia, however, has facilitated investment by Chinese hydropower developers. In December 2011, Cambodia inaugurated its first large hydropower dam, the 193 MW Kamchay Dam, developed as a BOT project by Sinohydro Corporation (Grimsditch, 2012) (see below). While lauded by the Cambodian government for increasing electricity supply and national energy security and reducing electricity prices, civil society groups have criticized the project for its construction in a national park, the limited public consultation and an overall lack of transparency in the decision-making process. In 2012, the Kirirom III hydroelectric station (18 MW) was also put into operation by a Chinese developer.

Four further large hydropower dams are presently under construction/ recently completed in BOT arrangements (see Table 7.2). Three are by Chinese companies in southwest Cambodia with a combined capacity of 684 MW. The fourth is being developed jointly between Cambodian, Chinese and Vietnamese companies. At least a further 11 medium to large hydropower projects hold MoU for feasibility studies; four are under study by Chinese companies, while six are allocated to South Korean companies and one to a company from Russia (Jona, 2011).

IPPs and BOT in Cambodia

The basic logic of IPP involvement in a BOT project is that investors carry the risk of the project's development, in exchange for a profit during project operation. Within this logic, governments and regulatory authorities should

Table 7.2 BOT hydropower projects in Cambodia in operation and under construction

Dam	Lead developer	Status
Kamchay Dam, Kampot Province (193 MW)	Sinohydro Corporation (China)	Inaugurated in 2011 44-year BOT contract US$280 million; financing from China Exim Bank
Kirirom III, Koh Kong Province (18 MW)	China Electric Power Technology Import–Export Corporation	Inaugurated in 2012 30-year BOT contract unknown cost; financing from China Exim Bank
Stung Atai, Pursat Province (100 MW)	Joint venture including Datang Corporation (China)	Inaugurated 2013 34-year BOT US$255 million; financier unknown
Lower Stung Russei Chrum, Koh Kong Province (338 MW)	Subsidary of China Huadian Corporation	Under construction; to be commissioned 2015 US$500 million; partly financed by China Exim Bank
Stung Tatai, Koh Kong Province (246 MW)	China National Heavy Machinery Corporation	Under construction; to be commissioned 2015 42-year BOT US$540 million; partly financed by China Exim Bank
Lower Sesan 2, Stung Treng Province (400 MW)	Royal Group (Cambodia) (39%) Hydrolancang International Energy Co (China) (51%) EVN International Joint Stock Company (Vietnam) (10%)	Under construction (reservoir clearance); to be commissioned 2017 30-year BOT US$816 million; financier unknown

Source: Grimsditch 2012; EAC 2013; note some company details vary with Jona (2011), Chea (2009) and Middleton (2008).

negotiate contracts to ensure risk is distributed as such. A conventional project financing model involves the IPP developer securing a contract to supply electricity to the utility, and then raising finance from private capital markets. In Cambodia, large IPPs, however, have negotiated to redistribute project risks to the government – and by proxy, to Cambodia's public, including electricity rate payers and tax payers. Ryder (2009, 54) argues that:

> To protect large IPPs against the risk of non-payment by EDC, the government has agreed to guarantee payment for electricity for the duration of their concessions. So if for any reason EDC cannot pay its IPPs, the government is on the hook.

Ryder (2009), citing an earlier World Bank (2006) study, suggests that an important reason why IPPs struggle to raise capital for projects in Cambodia is the poor financial management of EDC, the principal electricity buyer. Ryder (2009, 54) summarizes the World Bank (2006) study as showing that EDC 'has trouble collecting payment from government customers, it is vulnerable to political manipulation, it does not insist on competitive bidding, and it has no transparent price-setting mechanism in place to give investors' confidence they will be able to recover their costs and receive a fair return on their investment'. As of 2009, there was still no public process of bidding in place and investors were to submit their proposals directly to MIME (Oum, 2009), although bidding was reported for the Kamchay Dam in 2005 (Middleton, 2008, 56).

While reasonably comprehensive laws on environmental conservation and social protection exist on paper in Cambodia, overall implementation is weak and the institutional structure for water management in Cambodia is highly compartmentalized with limited coordination among key agencies (Chamreoun, 2006). While recognizing the need for and desirability of increasing the electricity supply, civil society groups in Cambodia have expressed their concern over how large hydropower projects are being developed with limited consultation with civil society and the public, lack of full regard for environmental and social protection, and a lack of transparency in the decision-making process. Civil society groups in Cambodia have sought to enforce existing laws on environmental and social safeguards, and to promote decentralized power and participatory power planning (Ryder, 2009). The government, meanwhile, has called for Cambodia's environment conservation goals to be balanced with the need to increase the electricity supply to ensure national economic growth (Reaksmey and Chen, 2011).

Kamchay Dam

In April 2005, the Cambodian government awarded Sinohydro Corporation, a Chinese SOE that is one of the world's largest hydropower companies, the right to develop the Kamchay hydropower scheme as a 44-year BOT project in Kampot Province, southwest Cambodia. The 193 MW project, costing US$280 million, with financing by the China Exim Bank, represented at the time the single largest investment by a Chinese company in Cambodia (Middleton, 2008). While the approval process, a financial risk guarantee from the government to Sinohydro Corporation, and the length of the BOT contract were all questioned by opposition lawmakers, supporters of the project from the ruling Cambodian People's Party justified the project on the basis that it helped meet Cambodia's need for affordable and reliable electricity (*Phnom Penh Post*, 2011). Full project construction commenced in 2007 and, as stated above, the project was inaugurated in December 2011. Financing for the dam was at first widely reported in the media as part of a massive US$600 million Chinese aid package in April 2006, but subsequent assessment reveals that the project is most probably funded by the China Exim Bank as a commercial loan (Grimsditch, 2012).

In July 2006, the government voted to provide a risk guarantee to Sinohydro; while 69 lawmakers voted in favour of the project, 10 abstained, with one stating that the project's contract had not been shared for the voting lawmakers to review and another questioning why the BOT contract was of such long duration (Middleton, 2008, 56). In its pursuit of the project, which Deputy Prime Minister H. E. Sok An referred to as 'a dream cherished by Cambodia since the 1960s' (*People's Daily Online*, 2006), the government has been willing to accept project financial risks, which it justifies as being necessary to attract the investment into Cambodia's power sector.

Meanwhile, a range of environmental and social impacts and risks have been externalized. The Kamchay Dam project is located wholly within Bokor National Park, flooding approximately 2,000 hectares of forest. The project itself required little resettlement, largely because settlement is not permitted inside the national park. However, at least 190 families living in the vicinity of the project who used to collect non-timber forest products before the project, including bamboo, rattan and mulva nuts, had their access to the forest dramatically reduced, resulting in a severe income loss (Grimsditch, 2012). They did not receive any compensation for their loss of livelihood, and to date alternative livelihood options have not been provided. Several durian plantations established illegally within the national park were also affected; in this case, compensation was provided for the loss of their crops but not for the loss of land, and the owners were generally satisfied with the compensation level (Grimsditch, 2012).

Several controversies erupted during the project's construction. Quarrying projected rocks that damaged people's homes and crops, although complaints were eventually resolved with the company (Grimsditch, 2012). The dam's construction also seriously affected the popular Touk Chhu tourist resort, a series of river rapids located just downstream of the national park and the Kamchay Dam's regulation dam (Middleton, 2008). River pollution caused by construction activities, increased garbage and discharge of untreated sewage resulted in a major reduction in tourism at Touk Chhu, with no compensation or livelihood replacement for the stallholders (Channyda, 2008). Furthermore, drinking water quality for residents in nearby Kamchay town, which draws its water supply from nearby the construction area, noticeably deteriorated during the construction period (Grimsditch, 2012).

At the time of project approval, local people knew little about the project, and there was still comparatively little community awareness as construction was almost completed in 2011 (Grimsditch, 2012; Middleton, 2008). Grimsditch (2012) observes that:

> Throughout the implementation of the Kamchay project's construction stages it is apparent the company and local authorities did not have a full plan for dealing with all of the project's impacts.

Gätke and Borin (2013, p. v) report that the project's environmental and social impact assessment (ESIA) was completed only in July 2012, and the project's environmental management plan in implementation is 'far below what is required based on the late ESIA'.

The Kamchay Dam, as Cambodia's first large BOT hydropower project, has set a precedent for future large hydropower projects. While civil society groups have sought to monitor large hydropower development in the country, a lack of transparency and accountability persists both by the relevant state agencies and the IPPs themselves. Under these conditions social and environmental impacts and risks are readily externalized by project developers onto affected communities. Even as domestic electricity generation contributes towards meeting Cambodia's growing electricity demand, the distribution of financial risks and returns between the project developers and the Cambodian government remains opaque, despite it being a matter of public interest.

Conclusion

The BOT/PPP model has now emerged as the principal investment vehicle for hydropower projects in both Laos and Cambodia. Introduced within a context of regional economic integration and partial liberalization across the wider Mekong Region, this policy shift towards the BOT/PPP model has been an important driver of transformation in water and energy governance in the region. While the growing role of the private sector since the early 1990s was first tempered by the 1997 Asian financial crisis, since the early 2000s, private sector investment has now taken a leading role in the planning and operation of hydropower dams in the Mekong Basin.

The divergent geographies, histories and political economies of Laos and Cambodia have resulted in different BOT/PPP policy models, revealing, unsurprisingly, that international policy diffusion articulates with local context despite the role of shared international policy actors including the ADB and World Bank. Laos has generally sought to take shareholdings within BOT projects via SOEs thus forming a PPP, while Cambodia has encouraged full private ownership of hydropower dams that would then sell electricity to EDC, the state-owned utility. Both countries, however, have received extensive policy advice and technical support from the World Bank and the ADB. In this sense, the BOT/PPP model has been subsidized and legitimized by the international financial institutions and Western bilateral donor agencies. At the same time, there have been strong proponents of the BOT/PPP model in each of the region's governments.

As BOT/PPP projects are negotiated both between power importing and exporting counties, and the host state and the IPP, inequalities emerge due to unequal access to various resources such as knowledge – for example, legal, scientific and project assessment – and to international financing. Furthermore, the risk that state agencies may pursue their own interests rather than that of the wider public good, and that individuals within the agencies may do the same, is very present. Discussions surrounding alternative forms of energy or energy savings thus become difficult. The 'take or pay' contracts[4] and long-term power purchase agreements, together with the risk guarantees and other concessions provided by governments and/or international financial institutions, ensure long-term revenue flows to project developers, and redistribute financial risks away from developers and on to the government (and thus the general public), electricity consumers, and affected communities. There is little evidence, for example, that the cost of decommissioning projects is seriously taken into account as governments negotiate BOT agreements, yet as the project ends up in the ownership of the government for the final stages of its operational life it is the government who must therefore shoulder this cost. Although this discounting is perhaps standard practice it can have significant financial repercussions as we are currently seeing in the United States with their drive to decommission outdated and inefficient dams.

The structure of the BOT/PPP model and the rush of private-sector investment in the Mekong Basin's hydropower development have created opportunities for project proponents to ignore or downplay the impacts of hydropower dams on people and the environment. Unsurprisingly, projects have focused heavily on profits and less on environmental and social mitigation. As demonstrated by the case studies, affected communities have been exposed to numerous environmental and social risks and impacts without their consent. This is compounded by weak government regulation in Cambodia and Laos and limited political space for civil society to meaningfully influence private-sector and government planning for water resources and electricity when it comes to the construction of large hydropower dams.

The growing role of the private sector via BOT and PPP in the provision of electricity has important implications for both water and electricity governance. The presence of influential IPPs with a preference for the construction of large hydropower dams (and other forms of large, centralized generation) in the absence of transparency, public participation and accountability in water and electricity planning and decision-making tends to certain biases. An example is the privileging of large-scale electricity options over alternatives that may be available, ranging from alternative generation technologies (including small-scale and renewable technologies) to energy efficiency and demand side management approaches, and balancing the multiple uses of river basins. While the BOT/IPP model has the potential to lever financing that is difficult for governments to access, it poses a unique water governance challenge between state and non-state actors. The notions of risk and uncertainty need to be further explored and shared by actors in a way that the commonly held goals of development – economic, social and environmental – are considered, and legitimate rights to access natural resources and social justice are ensured.

Acknowledgements

Naho Mirumachi would like to acknowledge funding from the British Academy Small Grants SG112581.

Notes

1 Whilst many of these projects are now under construction, they have proceeded led by different developers than those listed.
2 Larger dams are being planned, however, on the Mekong mainstream. Most notably, the Xayaburi Dam in Laos is the first hydropower dam to be constructed with a capacity of 1285 MW.
3 A consolidated licence 'is a licence, which may be the combination of some or all types of licences. The consolidated licence can be issued to EDC and to the isolated systems to grant the right to generate, transmit, dispatch, distribute and sale electric power to the consumers' (EAC, 2013).
4 A contractual arrangement that commits buyers to purchase electricity from IPPs whether there is demand for the electricity or not.

References

ADB (Asian Development Bank) (2003) *LAO PDR Power Sector Strategy Study*, Manila: Electrowatt-PA Consulting.

Bakker, K. (1999) 'The politics of hydropower: Developing the Mekong', *Political Geography*, 19: 209–32.

CDC (Council for the Development of Cambodia), CIB (Cambodia Investment Board), and CSEZB (Cambodia Special Economic Zone Board) (2014) 'Electricity' (webpage), www.cambodiainvestment.gov.kh/investors-information/infrastructure/electricity.html, accessed March 2014.

CDRI (Cambodia Development Resource Institute) and ANZ Royal Bank (2010) *Strengthening Key Sectors for Cambodia's Return to Growth, Sustainable Development and Poverty Reduction: Energy and Rail Infrastructure,* Cambodia Outlook Brief, 4, Phnom Penh: CDRI and ANZ Royal Bank.

Chamreoun, S. (2006) 'Scoping study of existing frameworks related to the World Commission on Dams strategic framework – Cambodia', in R. A. R. Oliver, P. Moore, and K. Lazarus (eds), *Mekong Region Water Resources Decision-Making, National Policy and Legal Frameworks vis-à-vis World Commission on Dams Strategic Priorities*, Bangkok and Gland: World Conservation Union (IUCN), pp. 7–14.

Channyda, C. (2008) 'Kampot tourism takes hit amid construction', *Cambodia Daily*, 24 March.

Chea, P. (2009) 'National power and hydropower development plans in Cambodia', paper presented at the MRC SEA Cambodia National Scoping Workshop for MRC SEA Hydropower on the Mekong Mainstream.

Cruz-del Rosario, T. (2011) *Opening Laos: The Nam Theun 2 Hydropower Project*, Singapore: National University of Singapore.

De, P., Samudram, M., and Moholkar, S. (2010) *Trends in National and Regional Investors Financing Crossborder Infrastructure Projects in Asia*, ADBI Working Paper 245, Tokyo: Asian Development Bank Institute.

Delang, C., and Toro, M. (2011) 'Hydropower-induced displacement and resettlement in the Lao PDR', *South East Asia Research*, 19(3): 567–94.

EAC (Electricity Authority of Cambodia) (2013) *Report on Power Sector of the Kingdom of Cambodia*, 2013 edn, Phnom Penh: EAC.

EPPO (Energy Policy and Planning Office) (2012) *Summary of the Thailand Power Development Plan 2012–2030 (PDP2010: Revision 3)*, Bangkok: EPPO, Ministry of Energy.

EPPO (2013) 'Energy statistics of 2013', retrieved March 2014 from www.eppo.go.th/info/cd-2013/index.html, accessed March 2014.

Forouzbakhsh, F., Hosseini, S. M. H., and Vakilian, M. (2007) 'An approach to the investment analysis of small and medium hydro-powerplants', *Energy Policy*, 35: 1013–24.

Gätke, P. and Borin, U. (2013) *The Kamchay Hydropower Dam: An Assessment of the Dam's Impacts on Local Communities and the Environment*, Phnom Penh: NGO Forum on Cambodia, Environment Program.

Goldman, M. (2005) *Imperial Nature: The World Bank and Struggles for Social Justice in the Age of Globalization*, New Haven, CT: Yale University Press.

Greacen, C. S., and Greacen, C. (2004) 'Thailand's electricity reforms: Privatization of benefits and socialization of costs and risks', *Pacific Affairs*, 77: 517–541.

Greacen, E., and Sukkamnoed, D. (2005) 'Laos: Did the World Bank fudge figures to justify Nam Theun 2?', *WRM Bulletin*, 95, www.wrm.org.uy/oldsite/bulletin/95/Laos.html, accessed Feb. 2014.

Grimsditch, M. (2012) *China's Investments in Hydropower in the Mekong Region: The Kamchay Hydropower Dam, Kampot, Cambodia*, Washington, DC: World Resources Institute.

Guttal, S. (2000) *Public Consultation and Participation in the Nam Theun 2 Hydroelectric Project in the Lao PDR*, Hanoi: Focus on the Global South.

Handley, P. (1997) 'A critical view of the build-operate-transfer privatisation process in Asia', *Asian Journal of Public Administration*, 19(2): 203–43.

Hirsch, P. (2002) 'Global norms, local compliance and the human rights–environment nexus: A case study of the Nam Theun II Dam in Laos', in L. Zarsky (ed.), *Human Rights and the Environment: Conflicts and Norms in a Globalising World*, London: Earthscan/James & James, pp. 147–71.

Hirsch, P. (2010) 'The changing political dynamics of dam building on the Mekong', *Water Alternatives*, 3: 312–23.

Houay Ho (2011) 'Houay Ho Power Company', www.houayho.com, accessed May 2011.

Huntington, S. P. (1965) *Political Order in Changing Societies*, New Haven, CT: Yale University Press.

International Rivers (2008) 'Power surge: The impacts of rapid dam development in Laos', www.internationalrivers.org/files/attached-files/intl_rivers_power_surge.pdf, accessed July 2011.

Izaguirre, A. K., and Rao, G. (2000) 'Private infrastructure: Private activity fell by 30 percent in 1999', World Bank Group, Private Sector and Infrastructure Network, Sept., http://rru.worldbank.org/ Documents/PublicPolicyJournal/215Izagu-10-20.pdf, accessed July 2011.

Jona, V. (2011) 'Cambodia energy status and its development', paper presented at the Cambodia Outlook Conference, Phnom Penh, 16 March.

Khamin, N. (2008) 'Case study nine: Houay Ho hydropower project', in S. Lawrence (ed.), *Power Surge: The Impacts of Rapid Dam Development in Laos*, Berkeley, CA: International Rivers, pp. 73–5.

Lamberts, D. (2008) 'Little impact, much damage: The consequences of Mekong River flow alterations for the Tonle Sap ecosystem', in M. Kummu, M. Keskinen, and O. Varis (ed.), *Modern Myths of the Mekong*, Helsinki: Helsinki University of Technology, pp. 3–18.

Lawrence, S. (2009) 'The Nam Theun 2 controversy and its lessons for Laos', in F. Molle, T. Foran, and M. Käkönen (eds), *Contested Waterscapes in the Mekong Region: Hydropower, Livelihoods and Governance*, London: Earthscan, pp. 81–110.

Levy, S. M. (1996) *Build, Operate, Transfer: Paving the Way for Tomorrow's Infrastructure*, New York: John Wiley & Sons.

Matthews, N. (2012) 'Water grabbing in the Mekong basin: An analysis of the winners and losers of Thailand's hydropower development in Lao PDR', *Water Alternatives*, 5: 392–411.

Middleton, C. (2008) *Cambodia's Hydropower Development and China's Involvement*, Berkeley, CA: International Rivers.

Middleton, C., Garcia, J., and Foran, T. (2009) 'Old and new hydropower players in the Mekong Region: Agendas and strategies', in F. Molle, T. Foran, and M. Käkönen (eds), *Contested Waterscapes in the Mekong Region: Hydropower, Livelihoods and Governance*, London: Earthscan, pp. 23–54.

Middleton, C., Grundy-Warr, C., and Yong, M. L. (2013) 'Neoliberalizing hydropower in the Mekong Basin: The political economy of partial enclosure', *Social Science Journal*, 43: 299–334.

MIME (Ministry of Industry Mines and Energy) (2003) *Preparing the Private Sector Assessment for the Kingdom of Cambodia: TA Report for the Asian Development Bank*, TA No. 4030-CAM, Phnom Penh: MIME.

Mirumachi, N. (2012) 'How domestic water policies influence international transboundary water development: A case study of Thailand', in J. Öjendal, S. Hansson, and S. Hellberg (eds), *Politics and Development in a Transboundary Watershed: The Case of the Lower Mekong Basin*, New York: Springer Verlag, pp. 83–100.

Mirumachi, N., and Torriti, J. (2012) 'Public participation and economic appraisal of the Nam Theun 2 Hydropower Project', *Energy Policy*, 47: 125–32.

Molle, F., Foran, T., and Floch, P. (2009) 'Introduction: Changing waterscapes in the Mekong region – historical background and context', in F. Molle, T. Foran, and M. Käkönen (eds), *Contested Waterscapes in the Mekong Region: Hydropower, Livelihoods and Governance*, London: Earthscan, pp. 1–19.

Morris, L. (1997) 'Eastern distributor faces Federal Court roadblock', *Sydney Morning Herald*, 10 Oct., p. 8.

Nguyen, A. T. (2012) 'A case study on power sector restructuring in Vietnam', paper presented at the Pacific Energy Summit, Hanoi.

Nguyen, T. D. (1999) *The Mekong River and the Struggle for Indochina: Water, War and Peace*, Westport, CT: Praeger.

NTPC (Nam Theun 2 Power Company) (n.d.) *The EGAT Power Purchase Agreement: Summary for Public Disclosure*, Nam Theun 2 Hydroelectric Project, Vientiane: NTPC.

NTPC (2005) *Environmental Assessment and Management Plan: Main Text*, Nam Theun 2 Hydroelectric Project, Vientiane: NTPC.

NTPC (2014a) 'Financing', www.namtheun2.com/about-ntpc/financing.html, accessed Feb. 2014.

NTPC (2014b) 'Shareholders', www.namtheun2.com/about-ntpc/shareholders.html, accessed Feb. 2014.

Oum, R. (2009) *Energy Sector Report: Cambodia*, Phnom Penh: British Embassy.

People's Daily Online (2006) 'Chinese firm to build hydroelectric dam in Cambodia', http://english.people.com.cn/200602(23)/eng20060223_245315.html, accessed March 2014.

Phnom Penh Post (2011) 'PM opens Kampot hydrodam', 8 Dec.

Poch, K., and Tuy, S. (2012) 'Cambodia's electricity sector in the context of regional electricity market integration', in Y. Wu, X. Shi, and F. Kimura (eds), *Energy Market Integration in East Asia: Theories, Electricity Sector and Subsidies*, Jakarta: ERIA, pp. 141–72.

Porter, I. C., and Shivakumar, J. (2011) *Doing a Dam Better: The Lao People's Democratic Republic and the Story of Nam Theun 2*, Washington, DC: World Bank.

Reaksmey, H., and Chen, D. H. (2011) 'Hun Sen inaugurates dam, criticizes NGOs', *Cambodia Daily*, 8 Dec. 2011.

REEEP Policy Database (2012) www.reegle.info/profiles/KH, accessed May 2014.

Ryder, G. (2009) *Powering 21st Century Cambodia with Decentralized Generation: A Primer for Rethinking Cambodia's Electricity Future*, Phnom Penh and Toronto: NGO Forum on Cambodia and Probe International.

Sayer, A. (1985) 'The difference that space makes', in D. Gregory and J. Urry (eds), *Social Relations and Spatial Structures*, London: Macmillan.

Singh, S. (2009) 'World Bank-directed development? Negotiating participation in the Nam Theun 2 Hydropower Project in Laos', *Development and Change*, 40(3): 487–507.

Sneddon, C., and Fox, C. (2012) 'Water, geopolitics, and economic development in the conceptualization of a region', *Eurasian Geography and Economics*, 53(1): 143–60.

Socialist Republic of Vietnam (2011) *Approval of the National Master Plan for Power Development for the 2011–2020 Period with the Vision to 2030*, 21 July, 1208/QD-TTg, Hanoi: Prime Minister.

Studwell, J. (2007) *Asian Godfathers: Money and Power in Hong Kong and Southeast Asia*, London: Profile Books.

Suhardiman, D., Silva, S., and Carew-Reid, J. (2011) *Policy Review and Institutional Analysis of the Hydropower Sector in Lao PDR, Cambodia and Vietnam*, Vientiane: IWMI, ICEM, and CGIAR Challenge Program on Water and Food.

White, W. C. (2008) *Evaluating Dam Sustainability: The Challenge of Lao Nam Theun 2*, Morrisville: Lulu.com.

WikiLeaks (2008) 'Laos: Plans for five large dams on the Mekong Mainstream Advance', www.wikileaks.org/plusd/cables/08VIENTIANE111_a.html, accessed Jan. 2013.

World Bank (2002) *Private Solutions for Infrastructure in Cambodia: A Country Framework Report*, Washington, DC: Public–Private Infrastructure Advisory Facility and the World Bank Group.

World Bank (2005) *Cambodia Power Sector: Technical Assistance for Capacity Building of the Electricity Authority of Cambodia*, ESMAP Technical Paper, 76 (Dec.), Washington, DC: World Bank.

World Bank (2006) *Cambodia: Energy Sector Strategy Review*, Issues Paper, Washington, DC: World Bank.

Wyatt, A. (2004) *Infrastructure Development and BOOT in Laos and Vietnam: A Case Study of Collective Action and Risk in Transitional Developing Economies*, Sydney: University of Sydney.

8 The politics of the Lower Sesan 2 Dam in Cambodia

Kimkong Ham, Samchan Hay and Thea Sok

Introduction

In Cambodia, hydropower development has been steadily increasing since 2000. Electricity demand and supply have been issues of contention in the country since 1979, however. As a result, the Royal Government of Cambodia (RGC) has paid considerable attention to planning new hydropower dams for energy production and also to rehabilitating existing dams (MIME, 2003; RGC, 2012a). In 2010, the total electricity supply in Cambodia was 2200 GWh or 2.2 million kWh. Approximately 58 per cent of this supply was derived from the country's installed capacity of 537 MW and the remainder was imported from neighbouring countries: 67 per cent from Vietnam, 32 per cent from Thailand and the remaining 1 per cent from Lao PDR. Due to increasing electricity demand, the amount of imported electricity increased by 48 per cent in 2010 compared with 2009 (Victor, 2011). In 2012, the Ministry of Industry, Mines and Energy (MIME)[1] stated that the total electricity demand for Cambodia was 1,000 MW, of which 400 MW was for Phnom Penh (Sokheng, 2012). This high reliance on imported electricity impacts Cambodia's energy security and is costly; and much of it is generated using diesel and heavy fuel oil (HFO) (World Bank, 2006). Electricity demand is expected to rapidly increase to 1,349 MW by 2015 and to 2,401 MW by 2020 (ECA, 2010).

In response to the increasing electricity demand, the RGC has established an energy policy and strategy (the 2006 Cambodia Energy Sector Strategy Review) to develop a series of domestic dams. These dams are expected to generate at least 50 per cent of Cambodia's electricity supply by 2020 (Victor, 2011). As part of this plan, several potential sites for both small and large hydropower dams have been identified, with four dams already operational. These four dams are: the Kirirom I (12 MW) completed in 2002; Kamchay hydropower (193 MW) completed in 2011; the Kirirom III (18 MW) completed in 2013; and the latest, the Lower Stung Russei Chhrum (338 MW), was put into operation in late 2013 (*DAP Newspaper*, 2014). There are currently three other hydropower plants under construction: Tatay (246 MW), Atay (120 MW) and the Lower Sesan 2 (400 MW) (Victor, 2011). The Lower Sesan 2 (LS2) Dam is perhaps the most controversial dam in Cambodia. It is considered a very important dam for

Cambodia's future energy needs and the expansion of hydropower across the country. Opposition to the dam, however, has come from environmentalists and advocacy groups, who say that it will significantly impact environment and socio-culture in the area. Further opposition has come from the communities who are concerned about the large number of people who would be resettled without proper consultation and compensation. The LS2 was approved in 2012 by the RGC, but at the time of writing, construction had not yet begun. When completed, it will be the largest dam in Cambodia. It is estimated that it will cost approximately US$800 million.

This chapter sets out to understand the energy debate and discussion in Cambodia, providing the context within which the LS2 will be developed. It then explores the dam itself, its history and the surrounding controversies. It then focuses on local communities who will be affected by the dam's construction, and how they have sought to challenge the dam builders, and secure adequate compensation and livelihoods. This study is based on research that employed a variety of methods: field observations, key informant interviews, focus group discussions and household interviews.

Current hydropower development in Cambodia

Electricity demands

Meeting rising electricity demands has challenged Cambodia's economic growth. According to MIME, in 2013 electricity demand increased at an annual growth rate of 15 per cent and the government needed to achieve an annual increase in generating capacity of 20 per cent to stabilize the electricity sector. The RGC has identified the development of the electricity sector as one of its four priority sectors (the other three being water, roads and human resources) (RGC, 2012b). The RGC reported that, in 2007, residential consumers used 49 per cent of the total electricity supply, followed by industry at 15 per cent, commercial users at 26 per cent, and the administrative sector, which used 10 per cent (RGC, 2009b). Studies indicate that electricity demand has increased due to a number of factors. First, Cambodia's population has grown by 14 per cent, from 11.4 million in 1998 to 13 million in 2008 (NIS, 1998, 2008). Associated with this population growth, Cambodia is experiencing rapid urbanization. Cambodia's average annual urban demographic growth rate rose by 4.34 per cent between 2000 and 2010 (World Bank, 2012, cited in CDRI, 2012).

Anticipating increasing electricity demands, the RGC issued the Power Sector Development Policy in 1994, which was designed to increase the electricity supply across the country (MIME, 2003). This policy was further integrated into the RGC's Rectangular Strategy Phases I and II, which focused on planned economic growth from 2006 to 2013. The strategic plans included the development of electricity generation, transmission lines and capacity as part of the government's effort to rehabilitate existing dams, and develop new dams and physical infrastructure throughout the country (RGC, 2006, 2009a). Increased access to the electricity grid has significantly contributed to increased demand. The number of people with access to electricity was less than 15 per cent of Cambodia's population in 2006, but this increased to 29 per cent in 2010 (World Bank, 2006). In 2010, most electricity consumption was in urban areas, where almost 100 per cent of households were electrified; the same was true of only 12.3 per cent of rural households (Victor, 2011). In 2010, electricity demand in Phnom Penh (population 1.5 million) grew by 25 per cent (Victor, 2011).

Economic growth has also contributed to increased electricity consumption. In 2006, the annual electricity demand per capita was only 48 kilowatt hours (kWh). This grew to 138.4 kWh in 2009 and 159.2 kWh in 2010 (Victor, 2011; World Bank, 2006). A key indicator of this economic growth has been the rapid rise in Cambodia's gross domestic product (GDP), which grew from approximately US$277 in 2000 to US$830 in 2010 (NIS, 2010). This growth has been fuelled by electricity-hungry industries, such as garment and footwear industries, hotels and restaurants, construction and the agricultural sector (CDRI, 2012). Industrialization grew at an average of 12 per cent per year between 2003 and 2008 (RGC, 2009a). The growth of the industrial sector is a significant burden on the existing electricity supplies.

Cambodia's current reliance on imported electricity has impacted prices and affordability. Cambodia's electricity price is among the highest in the world and significantly higher than other countries in the Mekong Region. Currently, the actual electricity price is based on utilization. The cost ranges from US$0.08 to US$0.15 per kWh for residential users, from US$0.12 to US$0.16 per kWh for commercial users and from US$0.12 to US$0.15 per kWh for industrial users (CDRI, 2012). The electricity tariff can reach up to US$ 0.18 per kWh in Phnom Penh, and up to US$0.30 per kWh in provincial towns and it varies between US$0.75 and US$1.25 per kWh in remote areas where electricity generation depends on small-scale private electricity producers (Middleton and Sam, 2008).

At present, the electricity supply is unreliable. There are frequent electricity blackouts in cities; and the high cost of electricity is a major disincentive to potential investors (Grimsditch, 2012). The RGC aims to 'increase the supply side while controlling the demand side'. By 2020, it aims to ensure that all villages in Cambodia will have access to electricity regardless of the source, and that by 2030, at least 70 per cent of Cambodia's people will have access to improved and reliable electricity grids (Victor, 2012).

Cambodia is a member of the Asian Development Bank's (ADB) Greater Mekong Subregion (GMS) programme, which envisages a regional electricity grid and significant levels of intra-regional energy trade. In 2008, the ADB reported that electricity demand among the six GMS countries grew by 9.8 per cent a year between 1990 and 2006. In Thailand and Vietnam, electricity production lagged behind consumption[2] (ADB, 2008). Cambodia, therefore, perceives energy markets among its neighbours, and has explicitly stated its intention to export electricity to them, claiming that this would be part of the GMS energy trade (MIME, 2013). Hydropower dams will certainly be the main foundation that will enable Cambodia to earn from its electricity sector.

An alternative way of managing electricity is to manage demand. Even though Cambodia's electricity consumption is still lower than most Association of Southeast Asian Nations (ASEAN) members, the RGC has introduced a variety of measures and initiatives to reduce consumption. It is working with other ASEAN countries to reduce energy consumption by at least by 8 per cent from its 2005 level by 2015 (ASEAN, 2012). It has attempted to do this by raising public awareness about electricity conservation, and cracking down on electricity theft (RGC, 2008). Cambodia also has an electricity tariff policy that is partially designed to be pro-poor; electricity consumption of less than 200 kWh is partially subsidized by the government (MIME, 2013). Recently, the RGC, with technical assistance from the European Union Energy Initiative Partnership Dialogue Facility (EUEI PDF), has been working on a policy, strategy and action plan of national energy efficiency to meet increasing energy demand. It is expected that energy demand can be reduced by 20 per cent by 2035 by focusing on efficiency in five areas, including distribution/transmission, end-consumer products, buildings/construction and industries (EUEI PDF, 2013).

Hydropower as a potential source of green and renewable electricity

In Cambodia, as in much of the Mekong Region, hydropower is generally thought of, and promoted by the government and its proponents, as a comparatively clean, low-cost and renewable energy source that relies on proven technology. It has low operating costs and a long operational life. In the past, it has been seen as a worthy energy source by countries that depend heavily on imported fossil fuels for power generation (WCD, 2000). This narrative was echoed by the RGC during the inauguration of the Kamchay Dam: 'the project provides power sources for stimulating the economic dynamics and ensuring growth, sustainable growth and social welfare by increasing the power supply capacity to serve the need of socio-economic activities' (RGC, 2011).

The government argues that the energy from the Kamchay Dam is useful for offsetting constantly rising oil, gas and coal prices, and other renewable energy sources (RGC, 2011). A 15-year Cambodian Energy Sector Strategy (2006 to 2020) has been established to promote hydropower energy independence through power trade and exchange with neighbouring countries, as well as regional integration by encouraging private investment (RGC, 2006). Furthermore, the pursuit of hydropower has been encouraged as part of the ASEAN Green Connectivity, which aims to develop 15 per cent renewable energy across the region by 2015 (ASEAN, 2012).

The RGC sees a number of potential benefits from hydropower. Hydropower is considered a solution to Cambodia's energy security problems, as it will relieve the country of its current reliance on expensive imported energy. Hydropower will also increase the country's use of renewable energy generation, in line with ASEAN regional goals (MIME, 2013). Cambodia's national report for the Rio+20 United Nations Conference on Sustainable Development 2012 indicated that the promotion of hydropower projects is a cornerstone of Cambodia's energy policy (RGC, 2012b).

There are significant obstacles to developing hydropower in Cambodia, however. The country has limited financial and capital investment to support hydropower and other energy alternatives (RGC, 2013a). To develop the energy sector and undertake large-scale dam projects, including the Kamchay and Lower Sesan 2 Dam, the RGC needs to secure investment from multiple private sectors locally, nationally and internationally. Small-scale hydropower projects and other renewable energy projects have not been attractive to targeted private investors (MIME, 2013).

The RGC expects that more than 50 per cent of the total electricity supply will be generated by hydropower projects by 2020. Cambodia expects to export more than 4,000 MW a year to neighbouring countries (Victor, 2011). As a result, about 60 possible sites for small and large hydropower dams have been identified across the country with the potential to generate 10,000 MW of electricity. Approximately 50 per cent of the electricity capacity will be generated from hydropower dams located on the Mekong River's mainstream and about 40 per cent of the capacity will be generated from dams on the Mekong's tributaries. The remaining 10 per cent will be located in the country's southwestern coastal area. In August 2013, MIME announced that three hydropower dams were operating, four were under construction and 13 were undergoing feasibility studies (Victor, 2011). With the number of dams rapidly increasing, the RGC acknowledges that potential social, economic and environmental impacts need to be taken into consideration (RGC, 2011). But, MIME argues, tributary dams will have fewer negative impacts on fisheries, ecological biodiversity and local livelihoods compared to mainstream dams. Furthermore, according to MIME, mainstream dams may be built in Cambodia, but not until 2030 (MIME, 2013). The RGC argues that 'we should not look at a single tree, we should look at the whole forest' and that 'the impact is a result of the natural consequences of economic progress and development, but no development can be done without having an impact on environmental and natural resources and human beings' (RGC, 2008). Civil society organizations (CSOs) in Cambodia generally agree that hydropower is a potential solution to the country's energy security; however, there is significant disagreement as to how to effectively plan and implement hydropower so that its social and environmental costs are addressed.

Despite MIME's predictions, a number of studies, including Baird (2009) and Baran *et al.* (2013), indicate that dam construction on the Sesan River (a Mekong tributary) will negatively impact the river's resources, including aquatic species and fisheries. These studies also indicate that the Yali Falls Dam (on the Sesan in Vietnam) is causing changes to water quantity and quality downstream. These concerns also extend to the Lower Sesan 2 Dam. An interviewee from Pluk village stated that the villagers are concerned about changes to the duration and timing of the annual floods that might lead to a reduction in available fish habitats, thereby affecting spawning grounds and habitats along the river which would result in lower fish production (Royal University of Phnom Penh, 2013).

Concerns over hydropower development

Hydropower development is controversial in Mekong countries, due to transboundary issues, ecological changes and its environmental and social impacts. The negative social and environmental issues associated with hydropower development have been critically addressed by concerned local and international NGOs, political parties and CSOs. CSOs play a role in ensuring that hydropower's negative impacts are minimized and acceptable to the affected communities. CSOs have complained that electricity should only be produced to meet domestic demand and that an accurate energy assessment needs to be undertaken (NGO Forum on Cambodia, 2012). These concerns contradict the RGC's intent to export electricity to neighbouring countries (MIME, 2013). It is not domestic electricity demand, but the ambitious goal of commercializing electricity through trade with neighbouring countries, which provides a big push in the implementation of the planned hydropower development projects at the expense of natural, environmental and social resources.

Concerns over the management of hydropower's impacts generally emerge in the EIA process. For instance, a variety of CSOs have argued that the EIA reports on dams, such as for the Kamchay and LS2 Dams, failed to adequately address their social and environmental costs and had limited public consultation.[3] Other criticisms of the dams by CSOs are that the government does not consider EIA and public consultation to be compulsory when starting such mega-projects. Some of these accusations are founded. For example, a full EIA for the Kamchay Dam was only completed in 2011, the same year that the dam began operation. In the case of the LS2 Dam, there is concern over the dam's design, which includes neither sediment nor fish passage features (Ryder, 2009). Changes to the dam's design were recommended by the Natural Heritage Institute (NHI) among other CSOs. NHI also recommended that the dam be redesigned to be smaller to reduce its potential social and environmental impacts (NHI, 2013).

The LS2's reservoir will flood 335 km² and displace an estimated 797 families with a population of approximately 4,500[4] from three communes (RGC, 2013a). The dam will also affect dozens of villages upstream and downstream along the Sesan, Srepok and Sekong Rivers (the so-called '3S Rivers') (PECC1, 2009). There are concerns about the loss of existing habitat for wildlife species in the Lumphat Wildlife Sanctuary and Virachey National Park because the Lower Sesan 2 Dam will flood buffer zones and some parts of those are protected areas. Moreover, fish biodiversity will significantly decrease because the dam will block fish migration routes (Baird, 2009). The catchments of the 3S Rivers cover just 10 per cent of the Mekong Basin's area. Nevertheless, 42 per cent of the Mekong's immense fisheries biodiversity is represented in these rivers, 89 of which are migratory species, and 17 species endemic to the system. The study also highlights that migratory fish account for 60 per cent of the total fish caught by fishermen in these areas (Baran *et al.*, 2013). Modelling studies suggest that the LS2 Dam will reduce the total fish biomass by 9.3 per cent in the whole Mekong River Basin (Ziv *et al.*, 2012) and around 6–8 per cent of the Mekong

River Basin's sediment and cause significant hydrological changes to the Tonle Sap Lake (Trandem, 2013).

Basing their work on the MRC's Basin Development Plan 2015–30 (BDP2) scenarios, which included between 16 and 78 tributary dams and up to 11 mainstream dams, Ziv *et al.* (2012) assessed different impacts on the biomass of migratory species and the risk of species extinction due to habitat loss. They found that the energy generated from all tributary dams (in the event that no mainstream dams were built) would be less, while the environmental risk would be higher than building six upper mainstream dams on the lower Mekong River (Ziv *et al.*, 2012).

The RGC, however, claims that the impact on fisheries is overstated and does not warrant a significant redesign of the dam. The government's solution for the blocked fish passage is to create a fish hatchery station to support fish raising in general and it has argued that fish can also be raised in the dam's reservoir. In addition to this, the government feels that Cambodia has many rivers that can supply fish and says it will encourage the development of aquaculture (TVK, 2011).

In an attempt to provide a more comprehensive compensation policy for the LS2 Dam, government officials and the provincial governor have stated that the new resettlement areas must be equipped with physical infrastructure such as schools, roads and public facilities before resettlement takes place. During fieldwork conducted for this study, it was observed that such preparations were under way, but progress was slow due to financial, material and human resources constraints. It was been reported that villages downstream of the LS2 Dam will not be provided with compensation; only a few families who have their land and farmland areas adjacent to the LS2 Dam construction site will be compensated.

Financing dam development

Financial investment on the Lower Sesan 2 Dam

The total investment cost of the LS2 Dam is US$781.52 million. The dam will be built under a 45-year build-operate-transfer (BOT) agreement including the five years of construction (RGC, 2013b). Discussions about the possibility of constructing the LS2 Dam started in 2006 between the RGC and Vietnam. In 2007, a feasibility study and full EIA were carried out under a Memorandum of Understanding (MoU) between Cambodia's MIME and Vietnam Electricity (EVN). The feasibility study was conducted by Vietnam's Power Engineering Consulting Joint Stock Company 1 (PECC1) in collaboration with relevant ministries of Cambodia, and the full EIA study was conducted by Key Consultants Cambodia (KCC), which is a national consultancy company, subcontracted from PECC1 (PECC1, 2009).

Initially, the dam was to be a joint venture between Cambodia's Royal Group (49 per cent of shares) and the Vietnam Electricity International Joint Stock

Company (EVNI) (51 per cent of shares) (Say, 2011). The Royal Group is a Cambodian business conglomerate best known for its investment in mobile phone networks, media, banking, insurance and trading (Royal Group, 2008). It has a banking arm, which includes ANZ Royal Bank, a venture with the Australia and New Zealand Banking Group Limited (ANZ). EVNI is a subsidiary of EVN, a state-owned enterprise engaging in construction, project management and hydropower projects (*Bloomberg Business Week*, 2013). In Cambodia, EVNI has been involved in a number of projects including the construction of transmission lines to connect the Lower Sesan 1, Sesan 5 and LS2 to Vietnam's power grid and the construction of the Lower Sesan 1 and Lower Sesan 5 (EVNI, 2013).

Progress on the project has been delayed for some time. In the communities that would be affected, it was rumoured that this was because of the dam's potentially severe negative impacts, and its controversial resettlement plan. The RGC, however, claimed that the delay was caused by financial problems within EVNI (MIME, 2013). In 2012, following negotiations, EVNI decided to reduce its shareholding to only 10 per cent to cover the cost of feasibility and EIA studies, while the remainder was acquired by the Royal Group with China's Hydrolancang International Energy Co. Ltd, as an additional partner under a new joint venture named Hydropower Lower Sesan 2 Co, Ltd (RGC, 2013b). It is not clear what the roles of each of the three players in the project will be; nor is it known how the project will be financed, although it has been reported that 30 per cent of project financing is derived from the shareholders' own internal resources, while the balance is derived from an undisclosed bank loan (70 per cent), most likely from China (Mekong Watch and 3S Rivers Protection Network, 2013).

The project was finally approved by the RGC with the signing of the Implementation Agreement and Power Purchasing Agreement in 2012. In 2013, the Cambodian National Assembly guaranteed the dam's payments (see Figure 8.1). During a visit to the kingdom, the Vice President of China Huaneng Group (CNG) (who own Hydro-Lancang International Energy) informed the Cambodian Prime Minister that construction would take three years, starting in 2014 and ending in 2017 (Reaksmey, 2014).

While EVNI's diminished role in the project was attributed to the company's financial difficulties, it occurred against a backdrop of increasing political and economic cooperation between the governments of Cambodia and China. Total investment by Chinese companies in Cambodia between 1994 to 2012 was reported to be US$9.1 billion (Kunmakara, 2013), while high-profile projects such as hydropower investments were publicly supported by senior government figures from the two countries (Middleton and Sam, 2008). Chinese hydropower investments in Cambodia are part of wider Chinese initiatives to identify new markets in low competition environments (Molle *et al.*, 2009).

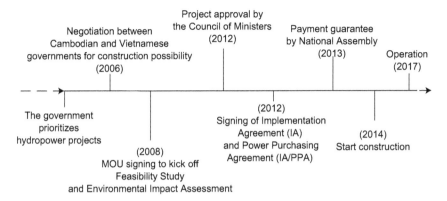

Figure 8.1 Process of LS2 Dam decision-making

Source: authors.

Lower Sesan 2 Dam's politics, impacts and changes in political support

The remaining part of this chapter will mainly discuss the LS2 Dam; in particular, the politics behind the narratives and the struggles of the CSOs and the local community will be uncovered. From the perspective of all CSOs in Cambodia, the LS2 is the worst dam ever to be built in the country. Moreover, the approval of this project has established an intense power struggle between the local community and the project owners/supporters.

Legal considerations surrounding resettlement and compensation

Before looking at the specific case of the LS2, it is worth looking at certain legal aspects with regard to the resettlement and compensation related regulations in Cambodia. Article 5 of the country's Land Law states that land ownership is protected and no one can be deprived of their property; however, this law allows the government to expropriate property in the public interest (RGC, 2001). In 2010, Cambodia adopted an Expropriation Law, which broadened the basis under which RGC could legally expropriate public and private property for projects that are deemed to serve the national and public interest (RGC, 2010b).

The RGC has stated that one of the aims of the Expropriation Law was to protect citizens by establishing processes and mechanisms for implementing expropriation. For instance, Article 22 states that expropriated properties will be compensated at market price when expropriation is declared. The price will be determined by an independent committee, which will be selected by the Expropriation Committee. The committee will be composed of representatives from the Ministry of Economy and Finance (MEF) and concerned ministries or institutions from the relevant provinces (RGC, 2010b). In Articles 19 and 34 of the law, however, the government can still expropriate property even if there remain unresolved disputes; property owners can continue to challenge the expropriation in court, or the compensation they received (RGC, 2010b).

Hydropower dam's compensation and resettlement policy

Cambodia has limited experience of large-scale dam development. Deciding upon 'fair and appropriate' compensation and resettlement is a key challenge; so too is the complexity associated with resettlement. For instance, the RGC argues that the Kamchay Dam required no resettlement because the dam's reservoir was located in Bokor National Park, where (officially) no one lived. Compensation was, however, given to some affected households whose agricultural land was affected by the construction of transmission lines and roads to access the dam construction site. Household surveys and interviews in the area indicate that the compensation procedure was based on agreeable negotiation between affected villagers and the dam builder and was facilitated by local authorities and the Department of Energy, Mines and Industry. For example, the loss of a mature durian tree was compensated at US$500 per tree because of its high value as a commercial cash crop, and agricultural land was compensated at US$3 per square metre.

Critics of the Kamchay Dam, however, have stated that the dam caused significant environmental problems and that the compensation and mitigation measures were inappropriate (see also Chapter 7 in this volume). Villagers whose houses and farms were hit by rocks from blasting at the dam site protested because they received no compensation (International Rivers, 2010). Furthermore, reduced access to non-timber forest products (NTFPs) and poor water quality, especially during the construction period, impacted tourism and livelihoods in the surrounding villages (Grimsditch, 2012). According to Grimsditch (2012), this was because the project's Environmental Management Plan was prepared just a few months before the construction was completed. There remains ongoing concern over how the compensation and the environmental impact mitigations are being implemented, including the restoration of project areas in Bokor National Park.

With regard to the LS2, compensation will include the loss of ecological environments and land due to flooding of the dam's reservoir. To address these issues, new compensation and resettlement policies have been established in the form of the Law on Guarantee of Payments. This new policy package includes concrete houses, residential and agricultural land, livelihood restoration activities and physical infrastructure such as hospitals, schools and rural roads (RGC, 2013b). The LS2's resettlement policy was initially proposed by EVNI, which classified a variety of house replacements and land allocations. The size of houses offered will depend on family size at the time of relocation (EVNI, 2011); and each household will receive 1.5 ha of agricultural land (EVN representative, interviewed on TVK, 2011).

Baird (2009) found that affected people at the LS2 site were dissatisfied with the resettlement and compensation package being offered because of the low quality of the houses and agricultural land allocation; moreover, the process could not meet even the minimum standards for public participation because not every affected upstream and downstream village was included in the EIA study (Baird, 2009). Our study also found that information on new locations and compensation was not effectively communicated to villagers because many did not speak or understand Khmer well. In response to the villagers' concerns, the RGC stated that every resettled household would receive an eight-member-size concrete house, 5,000 m² of residential land and an additional 5 ha of agricultural land (RGC, 2013b). In addition, the RGC stated that people would be allowed to select the new location of their houses. The dam's developers had agreed to compensate affected villagers between US$2 and US$44 for their fruit trees, depending on the species; and to build them three pagodas (Naren, 2014). In May 2013, however, field visits found that the search for land to resettle on continues. Despite these setbacks, the RGC claims that they are working to improve the compensation and resettlement process of the LS2 to make it a good model for future dams in Cambodia.[5]

The total budget for mitigation measures, compensation and resettlement for the LS2 Dam is said to be US$41.94 million (RGC, 2013b). The stated budget is understood to have been calculated based on a Sub-decree on Social Land Concessions (RGC, 2003), which is considered an improvement on previous policies. Interviewed villagers stated that most villagers had ongoing concerns about the dam's potential impacts on fisheries and how these impacts would affect their livelihoods and their ability to resettle. The upstream villagers who would be resettled due to the reservoir flooding stated that they did not want the concrete houses because they felt the houses were of lower quality construction compared to traditional wooden houses. Meanwhile, the downstream villagers complained that the dam would cause a loss of biodiversity resources, poor water quality and flooding. The government, however, argues that downstream villages will only suffer minor impacts.[6]

Our study found that most affected people in upstream and downstream villages were not aware of many details of the RGC's compensation and resettlement policy. With this ambiguity surrounding the compensation and resettlement implementation, many villagers were finding it difficult to continue their current livelihood activities as usual. The interruption in their daily lives has had significant impacts on food security and living standards. For example, most villagers are now hesitant about planting cash crops or maintaining their existing fruit crops and rice paddy because they may be required to relocate soon. The government has stated in accordance with its new policies that the affected people can only move to a new location when the physical infrastructure is ready; but while the start of construction date has been announced, new locations for resettlement have not been finalized and villagers have not been well informed or consulted. As a result, on 13 February 2014, the villagers living in the dam's reservoir lodged a petition to the Chinese Embassy in Phnom Penh, the government's ministries and the developer, demanding that negotiation be organized to discuss in detail the compensation and resettlement conditions.

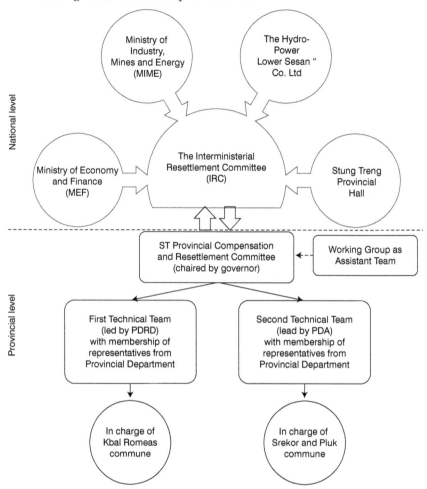

Figure 8.2 Structure of LS2 Dam's compensation/resettlement committee

Source: RUPP, 2013.

Compensation and resettlement mechanism

The compensation and resettlement plans for the LS2 appear to be clear and well prepared. They have been officially established at both the national and provincial levels (see Figure 8.2). At the national level, the Inter-Ministerial Resettlement Committee (IRC) consists of four key institutions: the Ministry of Economy and Finance (MEF) as chair, MIME, Hydro Power Lower Sesan 2 Co. Ltd and the Stung Treng provincial governor. This national committee works to approve resettlement action plans and compensation rates for any project requiring resettlement (Baird, 2009). At the provincial level, the committee is chaired by Stung Treng provincial's governor and supported by two technical teams led by relevant provincial departments. The provincial committee plays a significant role in the implementation of the resettlement and compensation plans. It communicates with the two technical teams, who engage with affected communities, map out potential resettlement sites, serve as liaison and receive complaints.

The establishment and implementation of the compensation and resettlement mechanism was assigned directly to the provincial level by the Office of the Council of Ministers; initial implementation was a challenge because the necessary resources were the responsibility of each of the lower implementing institutions. Even though the mechanism is in place, its enforcement is limited. The two technical teams have limited resources to produce any concrete results because of the uncertainty in the budget and a lack of necessary equipment.

Cultural and spiritual impacts

Cultural and spiritual impacts of the LS2 have not been included in compensation policy implementation. The cultural and spiritual impacts are deemed intangible and cannot be quantified in economic terms.[6] In Cambodia, minorities (i.e. non-Khmer) people live mainly in the northern provinces, including Kratie, Stung Treng and Ratanakiri. They have been described as the 'most ancient inhabitants of the land with strong cultural and economic ties to their environment' (Palmieri, 2010, 3). Culture plays a significant role as an arbitrator in the informal resolution of internal conflicts in these communities. Animistic worship and rituals are important elements in these cultures, and are believed to protect health, provide good fortune, increase agricultural production, improve livelihoods and sustain the economy. Livelihoods are based on animal husbandry, shifting cultivation and the collection of NTFPs from the natural forest.

Many of the people living in the area of the LS2 Dam are strongly united, and have deep respect for customary law, practices and religion (Phath and Sovathana, 2012). Villages both upstream and downstream of the LS2 Dam are populated by indigenous people including the Khmer Laotians, Phnong and Kroeng minorities, who profess a belief in Buddhism and/or animism.

As many as 88 per cent of the upstream villagers (out of 106 respondents) mentioned that their religion and tradition would be affected if they were relocated because their Buddhist temples, the guardian spirit of their village (*neakta*), the guardian spirit of the forest (*areak*) and their ancestors' graveyards would be flooded. In addition, the guardian spirit of the rivers (*neakta krahomkor*) would be affected. All of these spirits are celebrated from generation to generation, and are believed to protect the villagers from illness, bring them happiness and harmony, provide them with good businesses and agricultural activities, and protect them when they travel on the river.

An official from the Stung Treng provincial hall claimed that cultural losses had already been integrated into the compensation and resettlement plan. They are not, however, mentioned in the project's environmental management program (EMP) or in the Law on the Cambodian Government's

Guarantee of Payments to Hydro Power Lower Sesan 2 Co. Ltd. In any case, the local people have argued that the loss of their spirituality cannot be compensated.

Elders interviewed in the Srekor villages mentioned that these issues will have to be resolved traditionally if people are to feel more satisfied about the compensation package. Traditionally, the people in the areas surrounding the LS2 Dam site either bury or cremate their dead. Cremation occurs at the local temple, while burials are performed at communal graveyards close to the villages. These graveyards are extremely important spiritual sites. The families of the dead frequently pay their respects to the dead in order to attract good luck, make offerings of food and burn incense for them. They may invite Buddhist monks to perform ceremonies in the graveyards, especially on Phchum Ben Day (the day of the ancestors, an extremely important religious holiday in Cambodia) and Khmer New Year's Day. It is believed that the ancestors will be angry and curse them with illness or other problems if they fail to conduct these rituals. The elders stated that if the area were to be flooded, there would be two ways of dealing with the ancestors' graveyard: giving it up or moving it to the new location. Either way, spiritual and traditional rituals would have to be performed. This also applies to the movement of other spirits. Moreover, in seeking out a new place to live, local people must first ask the spirit of the land guardian (*neakta*) for permission by praying and through rituals.

Such respect and ceremony is considered vital to the villagers' lives and their harmony. For the people, any development or activity that displeases the spirits will harm them all. Without a concrete resolution, people's concerns will increase and they will seek an alternative solutions from outsiders, including political parties who might help to assuage their worries. Meanwhile, the affected people have been trying to voice their demands in public with support from CSOs. The demands from local people have also manifested in local elections, with political support shifting away from the Cambodian Peoples' Party (CPP) (the ruling party) in the commune/sangkat council elections. This change in support is explained in the next section that critically analyses how the local villagers have started to use their voice and rights to generalize and mobilize their demands.

Shifts in political support

The democratic process in Cambodia flourished after the Paris Peace Accords, signed by Cambodia's disputing parties in 1991. In Cambodia, there are two important elections: the National Assembly election and the commune/sangkat council elections. The first National Assembly Election was in 1993 and was managed by the United Nations Transitional Authority in Cambodia (UNTAC). This was the first time that the Cambodian people had been given the right to elect their political parties through a democratic process. The 1993 elections yielded a government with co-prime ministers, as a result of negotiation

Table 8.1 Timeline of political seats in the National Assembly

Year of election	Seat's political party in National Assembly		Remarks
	CPP	SRP	
1st mandate in 1993	51		Co-Prime Ministers from CPP and FUNCINPEC
SRP not yet formed			
2nd mandate in 1998	64	15	SRP got 15 seats
3rd mandate in 2003	73	24	SRP got 24 seats
4th mandate in 2008	90	26	SRP got 26 seats
5th mandate in 2013★	68	55	Cambodian National Rescue Party (CNRP) got 55 seats CNRP was formed through the merging of the SRP and Human Rights Party (HRP) in mid-2012

★ Preliminary result by NEC in Aug. 2013
Source: NEC, 2013.

between the National United Front for an Independent, Neutral, Peaceful, and Cooperative Cambodia (FUNCINPEC) and the Cambodian's People Party (CPP). Since the 1997 armed confrontation between the supporters of CPP and FUNCINPEC, the CPP has held the majority of seats in the National Assembly and the number of seats held by FUNCINPEC has continuously decreased. In 1998, the Sam Rainsy Party (SRP), which split from the FUNCINPEC party, became the largest opposition party in Cambodia, winning 15 seats in the National Assembly. The popularity of the SRP has been increasing from one election to the next (see Table 8.1).

The first commune/sangkat council election was held in February 2002. It aimed to establish and promote democratic development in Cambodia through decentralization and deconcentration. This election is held every five years to select the commune/sangkat chiefs and commune/sangkat councils. The ruling CPP currently administers most of the communes in Cambodia. Stung Treng province has a total of 34 communes. After the 2007 commune/sangkat elections, the CPP won the 34 commune chief positions, while the SRP secured only three positions as first deputy commune chiefs (NEC, 2007). After the 2012 commune/sangkat council election, however, the SRP won, for the first time, the position of commune chief in Srekor, which is one of the seven communes of the Sesan District of Stung Treng Province and one of three communes that will be flooded by the LS2 reservoir.

The local villagers indicated that the shift in political support was because they wanted to select a new commune chief who would be more responsive to issues in the commune, especially the dam project. The current commune chief

Table 8.2 Commune/sangkat council election result in LS2 Dam reservoir area

Name of commune	Commune chief position	
	2007	2012
Srekor (Upstream)	CPP	SRP
Kbal Romeas (Upstream)	CPP	CPP
Pluck (Downstream)	CPP	CPP

Source: NEC, 2007, 2012.

has been trying to help villagers strengthen their voices so that they can better negotiate the terms of compensation with the LS2 Dam company.

The villagers claimed that local authorities from the ruling party did not have the capacity and authority to effectively deal with some issues due to the government's administrative structure. Opposition parties campaigned on a platform to address unsolved problems with the dam and by promising impacted people that they would solve these problems if elected to office. The opposition parties went as far to say that they would stop the LS2 Dam project, reclaim community land allegedly abused by economic land concession companies, eliminate illegal logging, and allow local villagers to traditionally and freely collect NTFPs from natural forest areas.

If the project is built, the affected people want the local authorities to make sure that their livelihoods at the new locations will improve, and that they will get benefits from the dam. To achieve this, the people's concerns and demands must be carefully considered; however, the space for the people is limited, which causes them to seek alternative ways to convey their concerns and complaints to key players such as government institutions, the National Assembly and the dam company (Royal University of Phnom Penh, 2013).

The loss by the CPP in the commune/sangkat council election in Srekor commune is a message to the government regarding the LS2's development, and a reaction to the way mitigation measures are implemented. The new Srekor commune chief was seen to be very active in the public media, raising concerns over the issues of the LS2 Dam; his position does seem to have changed, however, from stopping the dam outright to making sure affected people will be better off.

Despite the political shift in the area, there has been no significant change in the government's position on the LS2 Dam: Stung Treng province has just one seat in the National Assembly, which is occupied by the CCP.

Conclusion

Hydropower dams are being constructed in Cambodia in order to meet growing electricity demands, and to reduce dependence on expensive imported electricity. The RGC has also, however, explicitly shown its intention to become an energy exporting country at the expense of natural, environmental and social resources. This ambitious goal has been operationalized and integrated into energy policy, strategy and plans, in which several hydropower dams have been built or rehabilitated and dozens more planned and studied, including the dams on the Mekong River mainstream, which have been a cause of grave concern among local and regional CSOs and local communities.

The construction of dams has been a controversial issue in Cambodia. At the same time, there is growing evidence of the impacts and flaws of the LS2 Dam and other dams in Cambodia elsewhere in the lower Mekong. Several studies have reiterated the severe negative impacts of dams including changes in ecology, changes in water quantity and quality, and social and economic impacts. In addition, the concerns of affected villagers are overwhelming and they have frequently taken to the streets demanding the cancellation of dam projects and bargaining for better compensation. There is no sign, however, that the RGC has changed its position on hydropower development in Cambodia. The government recognizes the possible negative impacts and is ready to sacrifice certain resources for electricity in the belief that it will eventually benefit the country as a whole. The question is, however, how well the impacts have been studied and understood.

In Cambodia, hydropower dam issues have been intensively argued among the government, opposition political parties and CSOs. Common problems exist around the quality of the feasibility and EIA studies and compensation plans. The RGC and dam investors have been requested to involve all affected villagers and to hold a meaningful process of public consultation in order to establish appropriate compensation. In the case of the LS2, lots of possible negative impacts have been predicted on the complex natural, environment and social resources. Despite this, spaces in which to deliberate these impacts between the state, the dam builders, communities and CSOs hardly exist. In addition, the compensation and resettlement plans contain flaws, and their implementation has been slow and uncertain due to the lack of finance and equipment.

The LS2 and the surrounding natural resources are being used and manipulated by several stakeholders, especially political parties, for their own benefit. To gain support, the opposition has focused solely on the negative impacts, while the government has focused only on the positive, and it is believed that these arguments will significantly influence coming local administration elections. Additionally, affected villagers were found have less trust in local government mechanisms; therefore, they depend heavily on the opposition party, CSOs and independent media to convey their messages to put to pressure on the decision-makers.

Notes

1 In early 2014, the MIME was split into two ministries – the Ministry of Mines and Energy, and the Ministry of Industry and Handicrafts.
2 In Thailand and Vietnam electricity consumption grew by 8.1 and 13.6 per cent a year respectively between 1990 and 2006; while electricity production grew by 7.3 and 12.6 per cent respectively.
3 Interview with an NGO representative, 6 March 2013.
4 The calculation of the total affected people is based on an average household size of six members (NGO Forum on Cambodia, 2012).
5 Interview with a representative of the Stung Treng provincial hall on 16 May 2013.
6 Interview with Stung Treng provincial officers, 16 May 2013.

References

ADB (Asian Development Bank) (2008) *Energy Sector in the Greater Mekong Subregion: Sector Study*, Manila: Operations Department, ADB.

ASEAN (Association of Southeast Asian Nations) (2012) 'Joint Ministerial Statement of the 30th ASEAN Ministers of Energy Meeting (AMEM)', www.asean.org/news/asean-statement-communiques/item/joint-ministerial-statement-of-the-30thasean-ministers-of-energy-meeting-amem, accessed Aug. 2013.

Baird, I. G. (2009) *Best Practices in Compensation and Resettlement for Large Dams: The Case of Planned Lower Sesan 2 Hydropower Project in Northeastern Cambodia*, Phnom Penh: Rivers Coalition of Cambodia.

Baran, E., Saray, S., Teoh, S. J., and Tran, T. C. (2013) 'Fish and fisheries in the Sekong, Sesan and Srepok Basins (3S Rivers, Mekong Watershed), with special reference to the Sesan River', in *MK3 Optimising Cascades of Hydropower: Fisheries and Environment*, Hanoi: International Center for Environmental Management.

Bloomberg Business Week (2013) 'Company overview of EVN International Joint Stock Company', http://investing.businessweek.com/research/stocks/private/snapshot.asp?prvcapid=51952372, accessed June 2013.

CDRI (Cambodia Development Resource Institute) (2012) *Annual Development Review 2011–2012*, Phnom Penh: CDRI.

DAP Newspaper (2014) 'Operating of Lower Stung Russei Chhrum Hydropower avoids Phnom Penh city lacking of electricity', www.dap-news.com/2011-06-14-02-39-55(80540)-2014-02-09-22-russeychrumhydro, *Deum Ampil News*, issued on 10 Feb., accessed Feb. 2014.

ECA (Economic Consulting Associates Ltd) (2010) 'The potential of regional power sector integration: Greater Mekong Subregion (GMS) transmission and trading case study', submitted to the Energy Sector Management Assistance Program (ESMAP) project on Regional Power System Integration (RPSI), London: EAC Ltd.

EUEI PDF (European Union Energy Initiative Partnership Dialogue Facility) (2013) *Development of an Energy Efficiency Policy, Strategy and Action Plan for the Kingdom of Cambodia*, Eschborn: EUEI PDF.

EVNI (Vietnam Electricity International) (2011) *Report on Construction Design Measures: Detailed Plan for the Resettlement/Resettled Cultivation Area – Lower Se San 2 Hydropower Project*, Phnom Penh: EVNI.

EVNI (Vietnam Electricity International) (2013) 'EVN International', www.evni.vn, accessed June 2013.

Grimsditch, M. (2012) 'China's investments in hydropower in the Mekong Region: The Kamchay Hydropower Dam, Kampot, Cambodia', www.bicusa.org/wp-content/uploads/2013(02)/Case+Study+-+China+Investments+in+Cambodia+FINAL+2.pdf, accessed Aug. 2013.

International Rivers (2010) 'Letter to SGS re Kamchay Hydroelectric BOT Project-Cambodia', www.internationalrivers.org/resources/letter-to-sgs-re-kamchay-hydroelectric-bot-project-cambodia-7395, accessed Nov. 2013.

Kunmakara, M. (2013) 'China to invest $9.6b in Cambodia', *Phnom Penh Post*, 1 Jan.

Mekong Watch and 3S Rivers Protection Network (2013) *Fact Sheet on Lower Sesan 2 Hydropower Project, Northeastern Cambodia*, Phnom Penh: Mekong Watch and 3S Rivers Protection Network.

Middleton, C., and Sam, C. (2008) *Cambodia's Hydropower Development and China's Involvement*, Phnom Penh: International Rivers and the Rivers Coalition in Cambodia.

MIME (Ministry of Industry, Mines and Energy) (2003) *National Sector Review of Hydropower Development*, Phnom Penh: MIME.

MIME (Ministry of Industry, Mines and Energy) (2013) Interview with a representative from MIME, 23 Aug. at the MIME.

Molle, F., Foran T., and Käkönen, M. (eds) (2009) *Contested Waterscapes in the Mekong Region: Hydropower, Livelihoods and Governance*, London: Earthscan.

Naren, K. (2014) 'Villagers want compensation for Lower Sesan 2 Dam construction', *Cambodia Daily Newspaper*, 14 Feb.

NEC (National Election Committee) (2007) *Official Result of the Commune/Sangkat Council Election in the 2nd Mandate in Stung Treng Province, Cambodia*, Phnom Penh: NEC.

NHI (Natural Heritage Institute) (2013) 'Lower Se San 2 Dam in Cambodia – Impacts and alternatives', presentation at Fishery Administration, 4 Sept., Phnom Penh.

NGO Forum on Cambodia (2012) *Lower Sesan 2 Hydropower Dam: Current Livelihoods of Local Communities*, Phnom Penh: NGO Forum on Cambodia.

NIS (National Institute of Statistics) (1998) *General Population Census of Cambodia 1998*, Phnom Penh: NIS.

NIS (National Institute of Statistics) (2008) *General Population Census of Cambodia 2008*, Phnom Penh: NIS.

NIS (National Institute of Statistics) (2010) 'National Institute of Statistics', www.nis.gov.kh, accessed May 2013.

Palmieri, S. (2010) *The Hidden Minorities: Representing Ethnic Minorities and Indigenous People in Cambodia*, Geneva: Inter-Parliamentary Union and the United Nations Development Program.

PECC1 (Power Engineering Constructing Joint Stock Company 1) (2009) 'Full environmental impact assessment of Lower Sesan 2 Hydropower', submitted to EVN International Joint Stock Company, Dec., Phnom Penh.

Phath, M., and Sovathana, S. (2012) *Kingdom of Cambodia: Country Technical Note on Indigenous Peoples' Issues*, Phnom Penh: International Fund for Agricultural Development and Asia Indigenous Peoples' Pact.

Reaksmey, H. (2014) 'Chinese push Sesan Dam talks with Hun Sen', *Cambodia Daily*, 2 April.

RGC (Royal Government of Cambodia) (2001) Land Law, Phnom Penh: RGC.

RGC (Royal Government of Cambodia) (2003) Sub Degree on Social Land Concessions, Phnom Penh: RGC.

RGC (Royal Government of Cambodia) (2006) *National Strategic Development Plan Phase I*, Phnom Penh: RGC.

RGC (Royal Government of Cambodia) (2008) 'The response of Cambodia Prime Minister to the opposition law maker', 8 Dec., Koh Kong.

RGC (Royal Government of Cambodia) (2009a) *National Strategic Development Plan Phase II*, Phnom Penh: RGC.

RGC (Royal Government of Cambodia) (2009b) 'Cambodian power development plan', presentation by MIME, 29–30 Sept., Phnom Penh.

RGC (Royal Government of Cambodia) (2010b) The Expropriation Law of the Kingdom of Cambodia, Phnom Penh: RGC.

RGC (Royal Government of Cambodia) (2011) 'Address of the Prime Minister of the Kingdom of Cambodia at the inauguration of Kamchay hydropower electric dam', 7 Nov., Kampot.

RGC (Royal Government of Cambodia), (2012a) *The Cambodian Government's Achievements and Future Direction in Sustainable Development*, Phnom Penh: RGC.

RGC (Royal Government of Cambodia) (2012b) 'Inauguration address of the Cambodian Prime Minister of the Kingdom of Cambodia at the opening of TATAIY hydropower plant', 18 May, Koh Kong.

RGC (Royal Government of Cambodia) (2013a) 'National Assembly over the approval of Lower Sesan 2 hydropower dam', debate on 15 Feb., Phnom Penh.

RGC (Royal Government of Cambodia) (2013b) Law on Cambodian Government Guarantee of Payments to Hydro Power Lower Sesan 2 Co., Ltd', Phnom Penh: RGC.

Royal Group (2008) 'Who we are', www.royalgroup.com.kh, accessed June 2013.

Royal University of Phnom Penh (RUPP) (2013) 'Field work report in Stung Treng province, Cambodia', 11–17 May, Phnom Penh: RUPP.

Ryder, G. (2009) 'Powering the 21st century: Cambodia with decentralized generation. A primer for rethinking Cambodia's electricity future', NGO Forum on Cambodia, Phnom Penh, and Probe International, Toronto, www.ngoforum.org.kh/docs/publications/HCRP_091023_Powering_Energ.pdf, accessed Sept. 2013.

Say, S. (2011) 'Construction nears for Sesan 2', *Phnom Penh Post*, 24 Jan.

Sokheng, V. (2012) 'Sustainable energy in Cambodia plans target by 2030', *Phnom Penh Post*, 14 Dec.

Trandem, A. (2013) 'China's overseas dam-building in the Mekong River Basin', presentation delivered to the Regional Public Forum: Mekong and 3S Hydropower Dams, 3–4 June 2013, Phnom Penh.

TVK (National Television of Kampuchea) (2011) 'Interview of fisheries administration in *Sesan Hydropower + Sea Fisheries* TV Documentary', Phnom Penh: TVK, available at www.youtube.com/watch?v=ni6jwdDhEbA, accessed Aug. 2013.

Victor, J. (2011) 'Cambodia energy status and its development', presentation at the Cambodia Outlook Conference, 16 March, Phnom Penh.

Victor, J. (2012) 'The energy efficiency challenge and opportunities in Cambodia', presentation to the 2nd EAS Energy Efficiency Conference, 31 July–1 Aug., Phnom Penh.

World Bank (2006) *Cambodia Energy Sector Strategy Review*, Washington, DC: World Bank, April.

WCD (World Commission on Dams) (2000) *Dams and Development: A New Framework for Decision-Making*, Report of the World Commission on Dams, London: Earthscan.

Ziv, G., Baran, E., Nam, S., Rodríguez-Iturbe, I., and Levin, S. A. (2012) 'Trading-off fish biodiversity, food security and hydropower in the Mekong River Basin', *Proceedings of the National Academy of Sciences*, 109(15): 5609–14.

9 Rethinking development narratives on hydropower in Vietnam

Nga Dao and Bui Lien Phuong

Introduction

Hydropower has been a key component of Vietnam's recent historical economic development. Since its independence in 1954, the Vietnamese state has paid increasing attention to developing multi-purpose reservoirs for electricity generation, irrigation and flood control (Dao, 2010).[1] This hydropower development reflects broader global trends of engineering-led water resource development (Reisner and Bates, 1990). It also reflects the Vietnamese state's dominant view on development, self-dependent, industrialization and modernization. Due to the 1954–75 war, Vietnam was only able to develop one dam in the north, the Thac Ba Dam. Although hydropower development was halted during the Vietnam War, it began again in earnest in the late 1970s with the construction of the Hoa Binh Dam, the second largest dam in contemporary Vietnam (1,980 MW). By early 2000s, cascade hydropower had been planned for all ten large river systems in the country (Dao, 2012). Overall, more than a thousand hydropower plants have been approved over the last two decades (MOIT, 2013).

However, alongside the benefits that hydropower has brought to the country's economy and developments such as electricity and irrigation, it has also caused significant impacts on local people and the environment in the dam's vicinity as well as in downstream areas. Over time, due to increased awareness of these problems, the public and some state perspectives on hydropower have shifted away from pro-hydropower positions to a more cautious approach.

This chapter critically analyses the narratives and decision-making process around hydropower development in Vietnam in general and in the Central Highlands in particular. It aims to contribute to a better understanding of Vietnam's hydropower development process over the last few decades, and further our understanding of the political and social dimensions of resource management in rural areas. The chapter draws on fieldwork conducted in Sa Binh and Ho Moong communes, Sa Thay district, Kon Tum province between December 2012 and May 2013. It is also informed by diverse data sources. These include: surveys with villagers; in-depth interviews with resettlement officers, authorities at different levels and village informants; legal documents and policy papers.

This chapter begins by providing an overview of hydropower politics and dominant narratives across Vietnam, followed by a discussion on hydropower decision-making. Special attention will be given to the shifting narratives, policies and power plays at various scales. The chapter then examines hydropower planning in the Central Highlands, focusing on the Sesan River and its two dams (the Yali Falls and Pleikrong Dams), and how narrative shifts have impacted hydropower practices in Vietnam and its Central Highlands.

Hydropower politics in Vietnam: whose narrative counts?

Hydro-social developments in Vietnam are not socially or politically neutral, but express and reconstitute physical, social, cultural, economic and political power relations (Dao, 2011). Hydropower development has played an important role in supporting the political agenda of the government and demonstrating its capacity for modernization and independence. At the inauguration of the Son La Hydropower Project, Prime Minister Nguyen Tan Dung (2012) said in his speech, 'With all our pride, we can say that the Son La hydropower project has continued the "The epic conquest of the Da River" and it has truly become a vivid expression of revolutionary heroism of a self-strong, self-reliant spirit that overcomes difficulties to rise up in the task of building of glorious Vietnamese Fatherland' (Nguyen Tan Dung, 2012).

Thus, dams play a critical role in the state-building process by symbolizing development. They are heralded as a prerequisite for industrialization, modernization and globalization, and as a way to improve the livelihoods of ethnic minority people. More than that, dam development has also been highlighted as a way to bring people together to unite the revolutionary spirit. This shows the government's will to lead Vietnam to develop on its own without relying on outsiders.

In fact, taming the Da and other rivers in Vietnam for the purpose of power generation has been an important objective of the Vietnamese state since its independence. On the Da River's mainstream, the three big dams are the Hoa Binh Dam (1,920 MW), the Son La Dam (2,400 MW – Vietnam's largest dam) and the Lai Chau Dam (1,200 MW). On the Da River's largest branch, the Nam Mu River, other dams are under construction, such as the Huoi Quang Dam (530 MW), and the Ban Chat Dam (220 MW). In addition, dozens of other medium-sized and small dams have been built or are planned on the Da River's tributaries. All ten large river systems in the country as well as hundreds of other smaller rivers have suffered from a similar density of dams over the last decade (Dao, 2012).

The ideals and nationalist fervour that dams have come to symbolize in Vietnam reveal themselves further in the case of the Hoa Binh Dam. For many people, especially government officials and developers, the negative impacts of big dams such as Hoa Binh are negligible compared to the benefits that the dams have brought and still bring, to the national economy (Nguyen, 2009). Since dams have become a symbol of development, local people need and should be willing to make sacrifices in the 'national interest' (Dao, 2011, 2012). In the 1980s and early 1990s, for example, postage stamps featuring images of the Hoa Binh Dam trumpeted a potent symbol of modernity. Stories created by the state to justify dam development in the uplands have become very popular. For a long time, people accepted this without question (Dao, 2010). People in urban areas and other parts of the country were happy to have electricity, and did not worry about the costs associated with electricity production (Dao, 2010). This may have been because, prior to 1992, little research had been conducted on dam impacts and there was no open debate allowed on the social and environmental costs. Thus, information about the negative impacts of dams as well as the true cost of building dams was not revealed for some time.

In the Central Highlands, the state uses policy statements emphasizing nationalism, modernization and development to justify hydropower development. These kinds of statements have strength and credibility at the countrywide scale because 'they are linked together, diffused and stabilized within narratives' (Fairhead and Leach, 1995, 1023). The narrative of hydropower in Vietnam, which emphasizes the positive impacts of the dams on the country's economy, has thus shaped the process of dam development in Vietnam since its independence.

One of the prevailing assumptions has been the need to help the highlanders to develop and move on from their 'backward way of life'. From the onset of the country's establishment in 1945, state leaders saw it as their responsibility to bring the highlanders into the independent Vietnamese nation-state (Dao, 2012). Even the 1992 Constitution highlights this:

> The state of the Socialist Republic of Vietnam is the unified state of all nationalities living together in the land of Vietnam ... The state implements the policy of all-sided development and step by step improves the material and spiritual life of ethnic minorities ...
>
> (Socialist Republic of Vietnam, 1992, 2)

One of the reasons for including the upland people in the state's development agenda was to improve or develop these mountainous areas so that they could contribute to national economic growth. Another large part of this state-assumed responsibility involved introducing the highlanders to a modern way of life, a strategy predicated on an ideological assumption that these people would progress over time from the state of *primitive highlander* to new *socialist man*, as a consequence of the modernizing policies of the state. Indeed, one of the justifications for the perceived need to develop ethnic minority groups was that these groups were judged to be at a lower level compared to the lowland Kinh in terms of both social and economic evolution. This development narrative has prevailed in Vietnam for several decades (Dao, 2012).

The state and investors consider dams to be not only evidence of modernity, but also engines to fuel the country's economic growth (Trang, 1995; Son La Peoples' Committee, 2006; Dao, 2009). According to a report by the National

Assembly's Committee on Science, Technology and Environment (MOIT, 2013), the Son La Dam provides 8 per cent of the country's total electricity output and has significantly contributed to the country's GDP. Annually, hydropower plants contribute VND 6,500 billion (US$3.25 million) to the treasury through various types of tax. Hydropower construction has therefore been credited as a positive factor for local economic restructuring, which also means that benefits from the dams are supposed to trickle down to the people.

In reality, however, the benefits and costs of dam construction are scaled, with benefits accruing at the national level, while costs accrue at local levels. By scaling benefits to the national level, the costs appear to be insignificant.

Dams are also justified as a way to improve the livelihoods of the ethnic minorities in the uplands of Vietnam and narrow the gaps among the various ethnic groups and between the uplands and the lowlands (Trang, 1995; Son La Peoples' Committee, 2006). This, however, raises an important political question about who has the right to determine development trajectories in Vietnam in general and in the upland areas in particular. In fact, while dams bring better infrastructure to the uplands in terms of roads and schools, they take away land for farming and flood forests, which in turn negatively impacts people's livelihoods. In the hydropower development process, upland people face many obstacles to making their voices heard, have very few opportunities to identify their own development needs and are obliged to follow the National Development Plan.

Dams in Vietnam are perceived by the central elites in terms of a vaguely conceived 'nation' that matters more than the local people who are disproportionately affected by these projects (Dao, 2011). This is everyday politics – the politics that concerns the control, allocation, production and use of resources, and the values and ideas underlying those activities (Kerkvliet, 2009). It amplifies the power difference between the lowlands and the uplands, and exaggerates the inequality among the different actors involved in development.

It is important to recognize that the civil society and media in Vietnam work differently from those in many other parts of the world. The first non-governmental organizations (NGOs) in Vietnam were established in the early 1990s by retired academics and government officials, to focus on small community development projects. Environmental and advocacy groups started their work mainly in the mid-2000s. The media in Vietnam are a state-owned industry and mostly serve the state's communication purposes. As a consequence, the voices of both NGOs and the media were almost absent from the hydropower planning process during the 1990s and early 2000s.

Overview of Vietnamese hydropower: its policy and approval procedures

During the first decade of the twenty-first century, Vietnam experienced an average annual economic growth rate of 7–8 per cent, and demand for electricity from the industrial sector and for domestic consumption increased rapidly (Dao, 2011). According to the Vietnamese Government's socio-economic development strategies for the period 2011–20, the government seeks to maintain Vietnam's economic growth at 7–8 per cent per annum (Government Portal, 2013). Vietnam's GDP in 2020 is expected to be 2.2 times that of 2010; GDP per capita will reach US$3,000. In addition, the industrial and service sectors are planned to constitute about 85 per cent of the total GDP, which results in more demand for electricity (Government Portal, 2013). As hydropower is considered by the government to be one of the cheapest forms of electricity production in the country, which can significantly contribute to meeting the country's demand for energy, hydropower projects are omnipresent in Vietnam. This is emphasized by a current joke that states that 'everyone is involved in hydropower and every household is building a dam'. Since 2003, the MOIT and Provincial People's Committees have directed hydropower planning for hundreds of rivers nationwide. Between 1993 and 2013, the government, the Prime Minister and the ministries issued at least 50 decrees, decisions and circulars to guide the implementation of policies and laws related to hydropower development (MOIT, 2013). The National Power Master Plans approved by the Prime Minister in 2007 and 2011 have prioritized the development of hydropower sources, particularly multi-purpose projects. The Power Development Plan approved in 2011 aimed at raising hydropower output from 9,200 MW to 17,400 MW by 2020 (Power Development Plan No. 7, 2011). As a result, even though the energy sector aims to reduce the hydropower ratio from around 40 per cent in 2005 to 21.3 per cent by 2020 (Power Development Plan No. 7, 2011), hydropower currently contributes 48 per cent of the total energy production in Vietnam (MOIT, 2013).

According to Vietnam Electricity (EVN), hydropower plans have been approved by the Prime Minster, the MOIT and the Provincial People's Committees (including cascade hydropower, and national and provincial small hydropower plans) for a total of 1,237 hydropower projects, with an installed capacity of 25,968.8 MW (MOIT, 2013). These plans have been approved by agencies for construction with the aim of achieving the goals of industrialization, modernization and socio-economic development. For hydropower cascades on large rivers and their tributaries with a capacity of at least 30 MW, the MOIT has appointed EVN to study, appraise and submit hydropower plans to the Prime Minister or relevant government authority for approval. As a consequence, between 2000 and 2013, a nationwide total of 899 hydropower projects with an installed capacity of 24,880 MW was in the pipeline (MOIT, 2013). This rapid hydropower development raises many questions that are relevant to this chapter. How are decisions made? Who makes them? What is prioritized during these decisions? We will examine these questions in the following sections.

Dam planning and approval procedures

In order to support its energy development plan, the government's Decision No. 66(2006)/QH11 on electricity investment divides energy projects into those at the national level and those at the provincial level. This classification helps to define who has the decision-making authority depending on the type of project. The approval of a hydropower project with an installed capacity greater than 30 MW is made by the Prime Minister while projects with an installed capacity of less than 30 MW are decided upon by Provincial People's Committees. The decision to proceed on the Hoa Binh and Son La Dams was made by the National Assembly.

Hydropower projects are not cheap to construct – approximately 25 billion VND (US$1.25 million) per MW. As a result of these costs, private investors do not usually invest in large projects.[2] Foreign investors do not invest in hydropower in Vietnam because the state subsidizes the energy sector and regulates the selling price. As a result, the majority of these types of projects (greater than 30 MW) are implemented by state-owned companies.

According to state policy, both state and non-state investors are only allowed to start their projects after having the projects appraised. Decisions on dams are supposed to be made only after consideration of their scale and impact. EIA reporting was made mandatory in the 1993 Law on Environmental Protection. EIAs must be conducted together with feasibility studies, and all potential negative impacts of any project or programme must be ameliorated by the adoption of mitigation plans. Investors have to carry out (by themselves or using consulting services) EIAs and submit these for approval to the Ministry of Natural Resources and Environment (MONRE). MONRE will then establish an Appraisal Committee to review the EIA. Before granting approval, MONRE sends the submitted EIA report to other relevant ministries (such as the Ministry of Agriculture and Rural Development, MARD) to have them reviewed with regard to relevant issues such as impacts on forests or biodiversity, resettlement, etc. Smaller projects (less than 30 MW) are usually approved at the provincial level without EIAs. Investors in such projects are mainly from the private sector. Regardless of a project's size, if it impacts national parks or reserves, it has to have an EIA and be approved by MONRE.

The reality on the ground is somewhat different. If state-funded projects are included in the national plan, they sometimes start before procedures such as EIAs or project appraisals are completed. In the meantime, non-state-funded projects usually have to follow all of the procedures required by law and their reports are usually reviewed more strictly. Once a state-funded hydropower project falls under the national strategic plan, other procedures are largely formalities. An inadequate EIA report need not result in a project being rejected. The appraisal board (presided over by MONRE) will issue instructions to continue rewriting or amending the report until it is approved. This explains why the final EIA report for the Son La Dam was only issued in 2007, four years after its construction started.

While provincial authorities are key actors at various levels, they are provided with very few opportunities to comment on nationally sanctioned hydropower projects in their provinces. If there are complaints and petitions, these have very little influence if the project has already been approved.

At the national level, consultation has not been much better. According to some interviewees at ministerial levels, even during the late 1990s and early 2000s, consultation with ministries, departments and agencies for national hydropower planning remained ineffective and sometimes superficial. When the plans reach specialized departments and agencies, a few officers are assigned to read them and comment. However, they might not bother to comment if they know that the project has already been approved, because such opinions would be pointless.

As mentioned above, civil society has not played a significant role in the hydropower planning process. Its participation in individual project appraisals has also been limited. While civil society has never managed to stop the construction of a dam in the national plan, it may be able to influence dam design. The government listened to suggestions from the Vietnam Union of Science and Technologies Associations (VUSTA) to reduce the size of the Son La Dam from 3,600 MW to 2,400 MW.[3] Even though the Son La Dam was the only case during the 1990s for which the government was under pressure to change its decision related to hydropower, it provided an incentive for civil society to develop a stronger movement, which has challenged the government's hydropower narrative over the last 15 years.

The emerging critical hydropower narrative

The importance of hydropower to the government's ideology of modernization and economic growth prevented meaningful research into its impacts until the early 1990s. In 1992, a few international and national scholars began to challenge the dominant hydropower narrative by analysing its social and environmental impacts (Hirsch *et al.*, 1992; Diep, 1992, 1997; Trang, 1995). These studies were the first to highlight the social and environmental impacts of the Hoa Binh Dam, which had been the pride of Vietnam's development strategies for two decades. People outside of the dam site started to learn about the cost of having electricity for modernization and economic growth. However, during the 1990s, this knowledge was mostly limited to academics.

In the Central Highlands, during the 1990s, state policies prevented the emergence of these critical voices. By wrapping hydropower in nationalistic ideologies and stating that it was key to economic growth, the government left little room for alternative views or counter-arguments. This began to change among the wider public in the early 2000s, when the true costs of dam development began to be revealed. Since 2002, despite the political and sensitive nature of state-owned dam-induced resettlement, scientists and environmental organizations have begun to do research on this topic (CRES, 2001). NGOs and social movements in Vietnam have recently emerged as a strong and influential voice, especially since the establishment of the Vietnam Rivers Network (VRN) in 2005 under the coordination of another NGO, the Centre for Water Resources Conservation and Development (WARECOD). Dam-related issues have become more open to the public. Academics and NGOs have worked together to conduct studies on the impacts of dams all over the country. The combination of both academic and NGO critiques and research into hydropower development has resulted in an emerging critical counter-narrative.

This new counter-narrative has created a dialogue between the state and NGOs in relation to hydropower development. The state constructs an image of dams as a symbol of development and fuel for national economic growth, while some NGOs, media and academics question the problems arising from the rapid promotion of dam building. Even though it is not easy for networks to operate in Vietnam,[4] VRN has been able to mobilize a number of scholars to do research on dam issues, including: VUSTA, 2006; Doan and Nguyen, 2006; Duong, 2009; Vo, 2006; Hoang and Vo, 2006; Hoang, 2006; and Chu *et al.*, 2007 (VRN, 2011). These studies by VRN's members show how dams have caused negative impacts on the environment and local people. The studies also highlight uneven development between the uplands and lowlands, and show that the people who have suffered in all of the dam projects are mostly ethnic minority people – or people marginalized in peripheral areas.

Scientists and activists have shared their study results extensively in the VRN's annual meetings and workshops at the provincial level, in Hanoi, and have used the results for lobbying and advocacy purposes. The VRN works closely with other regional and international networks and organizations that engage in dam-related issues. They have provided the public and affected people with comprehensive information about the impacts of different dams on the environment and on local people. They have also organized campaigns against dam planning in different locations such as Lao Cai and Dong Nai. High-quality research emerging from the network has helped the government to understand the important role of the VRN in the protection of Vietnam's rivers. As will be examined below, this has contributed to the adjustment of policies on dam building, resettlement, compensation and rehabilitation. And at the same time, it has raised awareness among the public and affected people about the need to pursue more equality in development programmes. A number of environmental journalists have joined the VRN and become actively involved in river protection work.

As a result, media outlets, such as television channels and newspapers, have been active in covering this controversial topic.[5] Since 2008, dam topics have been widely discussed in the media. While there was almost no media coverage of the impacts of the Hoa Binh or Yali Falls Dams during the 1990s, the problems associated with the Pleikrong and other dams made the headlines in various newspapers, for instance, 'Pleikrong Dam – mistakes everywhere' (Baomoi, 2009), 'Hydropower in the Central Highlands and its associated problems – deforestation for hydropower', 'Dry river – thirsty people', 'Gave land to hydropower and went hungry' (PhapLuatTPHCM, 2011a, 2011b, 2011c); 'Hundreds of families got into difficult situations due to hydropower' (CAND Portal, 2012); and 'Medium and small hydropower in the Central Highlands – Stop before it will be too late' (VoV, 2013), 'Hydropower kills rivers' (Thanhnienonline, 2013), etc.

The fact that the media have been able to report on these issues is a significant development in the political landscape of hydropower within Vietnam. The media's freedom to critique a state company for its failures in the development process sends a warning to other developers that their mistakes will be exposed. This exposure resulted in the MOIT forming an inspection team to investigate the Pleikrong Dam in 2009 (*Dai Doan Ket*, 6 September 2009). The impacts of dams, therefore, have become known to people outside of the dam's vicinity, which has helped more people to become aware of the issues and raise their concerns. In addition, more and more NGOs and researchers are doing research on dam issues. All of these changes play a significant role in changing people's perspectives on hydropower and allowing the politics of energy development to be openly debated and contested.

Hydropower in the Central Highlands: a contested driver for economic growth

The Central Highlands, with their rich biodiversity and culture and special geographical position, have always been considered by the Vietnamese state as a strategic region of the country. Identified as one of the seven important ecological economic zones of the country, the Central Highlands consists of five provinces: Kon Tum, Gia Lai, Dak Lak, Dak Nong and Lam Dong, with a total area of $54,474 \, km^2$, constituting 16.8 per cent of Vietnam's total area. The region had a population of 5,107,437 in 2011 (Vietrade, 2011). A dense network of rivers, many of which flow through steep gorges, makes the Central Highlands extremely attractive from a hydropower development perspective. There are four main river systems in Vietnam originating in the Central Highlands. The Sesan River in Kon Tum province and the Srepok River in Dak Lak province are both Mekong River tributaries. The Ba-Ayun River in Gia Lai merges with the Da Rang River, and the Dong Nai River in Dak Nong and Lam Dong both flow to the East Sea. The hydropower potential of these four river systems accounts for 25 per cent of the country's total hydropower potential, and could theoretically generate 15–16 billion kWh of electricity per annum (Vietrade, 2012). Dams such as the Yali Falls Dam, the Se San 3, 4, and 4a, the Pleikrong, the Upper Kon Tum, etc., have all had both negative ecological and social impacts and economic benefits.

The area is populated. The mountains and valleys of the region are home to 47 ethnic groups and a diverse system of fauna and flora (Vietrade, 2011). The Jarai group is the largest (350,776 people), followed by the Ede (306,333 people), the Bahnar (190,259 people), the Sedang (140,445), the Hre (120,251), the Mnong (104,312), the Jeh (31,343), etc. (CEMA, 2011). Among the minorities, two main languages are spoken: Mon Khmer and Chrau (Hickey, 1982). The area borders Lao PDR and Cambodia in the west and southwest. During the war, this area received much attention from both sides because various resistance forces were based there (Hickey, 1982). After the war, the state and Communist Party promulgated many policies in the highlands under the pretence of economic development, namely: (1) migration programmes that moved large numbers of people from overcrowded northern provinces to the Central Highlands; and (2) the development of hydropower cascades on the region's main rivers. The dominant narratives for both programmes were that they would help boost the region's socio-economic development. However, these programmes significantly affected the Central Highlands' population structures as well its development trends.

Many scholars have argued that these migration policies are part of a broader state policy and agenda to redistribute the population while strengthening government control over the ethnically diverse and geographically peripheral upland regions (Hickey, 1982; Hardy, 2002; Salemink, 2003). After the war ended in 1975, in an attempt to integrate the Central Highlands into the new socialist state, the government facilitated a new programme that encouraged people to migrate from the lowlands to the Central Highlands. In the late 1970s, a large number of lowland Vietnamese were resettled in upland economic zones (Hickey, 1982). There was subsequently a wave of spontaneous migration to the Central Highlands in the 1990s (Salemink, 2003).[6] The population structure of the Central Highlands has, as a consequence, changed dramatically. Between1976 to 2010, the proportion of ethnic minorities in the population of the Central Highlands declined from 70 per cent to 25 per cent. By 2010, about 70 per cent of the Central Highlands' population were lowland Kinh (Vietrade, 2011), pushing indigenous minorities deeper into the forests (Salemink, 2003).

If migration has significantly altered the population structure of the highlands, the development of hydropower cascades in the Central Highlands has transformed their social and agrarian landscape. Over the last two decades, hydropower development in the Central Highlands has displaced thousands of people, and submerged many towns and villages and thousands of hectares of farming land and forests (MOIT, 2013). The local people's limited access to land caused by the migration programme has been critically exacerbated by hydropower development. Thus, both programmes, while having similar narratives of promoting socio-economic development, have significantly affected the lives and livelihoods of ethnic minorities in the region and their access to land and other resources. Impacts of the two programmes, on the one hand, are intertwined, reinforcing each other. On the other hand, they are contributing factors to the changing narratives of hydropower development in the Central Highlands and in Vietnam as a whole.

Since 2003, a series of new hydropower projects has been planned and developed in the Central Highlands. Up until the 1990s, the Yali Falls Dam (720 MW) was the only large plant that caused resettlement in the Central Highlands.[7] In the 2000s, the situation changed dramatically. Coinciding with government policies to promote hydropower development, the development of projects in the Central Highlands accelerated rapidly: the Pleikrong (100 MW), Buon Kuop (280 MW), Sesan 3 (250 MW), Sesan 3A (108 MW) and many smaller projects were constructed. These dams were primarily built along the Srepok and Sesan Rivers. By July 2013, there were 118 completed hydropower projects of all sizes, with a total installed capacity of 5,798 MW, and another 75 projects were under construction in the Central Highlands. There were also around 200 other projects under study and/or about to begin construction (Vnexpress, 2013).

This massive hydropower expansion has been important to meeting Vietnam's growing energy demands, but it has not been without negative impacts. Hundreds of thousands of people have been relocated to make way for the dams. Unlike the migration programme that has altered the region's population structure, dam displacement has transformed local landscapes and people's livelihoods. The majority of the resettled communities had to move away from the river and the area of land that they could farm was significantly reduced, on average from 5–7 ha to around 1 ha per household (Dao, 2012). Dams not only flooded farmland and houses, they also submerged thousands of hectares of forest. In the Central Highlands alone, dams caused deforestation of 22,770 ha (MOIT, 2013). In the next section, we will take a closer look at one catchment in the Central Highlands, the Sesan River, to examine how power is exercised in hydropower development, the politics surrounding it and how hydropower narratives have changed.

The Sesan River Basin

The Sesan River and its main tributaries, the Dak Bla and Krong Poko Rivers, drain part of the Central Highlands of Vietnam. The river flows in a southwesterly direction through Kon Tum and Gia Lai provinces in Vietnam, and then joins the Srepok River in Cambodia before merging with the Mekong River. It has a total catchment area of 18,570 km², of which some 60 per cent (11,450 km²) is located in Vietnam. The length of the river is about 462 km, of which 210 km are in Vietnam (National Hydropower Plan Study, 1998).

The Sesan River has the third largest hydropower potential in Vietnam (2,000 MW), after the Da River (6,800 MW) and the Dong Nai River (2,900 MW) (Nguyen, 2009). Hydropower cascade planning on the mainstream of the Sesan River was approved by the government, to include six plants with a total design capacity of 1,738 MW and an average annual electricity production of 8,300 GW. The Sesan catchment master plan highlighted the potential benefits that the hydropower projects would bring to the area in terms of economic, social and environmental development (National Hydropower Plan Study, 1998). The plan estimated that hydropower projects would create revenues of US$430 million per annum, attracting thousands of labourers from both outside and inside the catchment during the construction and subsequent operation. Thus, the Sesan has the second highest density of hydropower cascades after the Dong Nai. Most of these dams are located in Kon Tum province (see Figure 9.1), and will have a total storage capacity of approximately 3.7 billion cubic metres. The Sesan catchment will soon be turned into a very large reservoir with a length of about 150 km from the Upper Kon Tum Dam to the Sesan 4 Dam. Supporters of this scheme, who do not reside in the vicinity of the dams, consider this

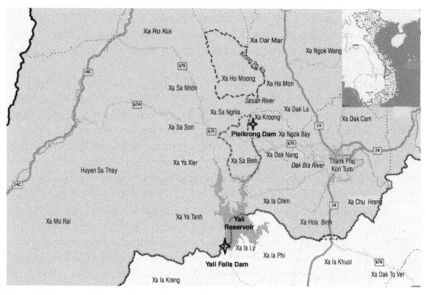

Figure 9.1 Map of the study sites for this chapter

change positive. As highlighted in the news, 'the dams will not only create more stable climate conditions for the region, and help improve farming conditions and arable land, but they will also enable favourable conditions for aquaculture, irrigation for farming in the surrounding areas' (Vietrade, 2012).

It is worth noting that most of the large dams on the Sesan River are huddled in one district of Kon Tum province – Sa Thay district. When asked about these dams, most district authorities said that they were happy about the dams because they had contributed to the district's infrastructural and socio-economic development. Only one official from the Division of Ethnic Minorities said: 'the dams have caused a lot of problems. They have destroyed the landscape and caused serious land loss. Many ethnic minority people have had to go deeper into the forest to clear land for farming and the land is still not enough; dealing with farming land shortages and other social problems in resettlement sites has become a burden for the local authorities.'[8] In the following sections, we will discuss the Yali Falls and Pleikrong Dams to illustrate changes in the decision-making process and narratives.

Yali Falls Dam

The Yali Fall Dam was planned to be the second largest dam in Vietnam, after the Hoa Binh Dam (see Box 9.1). The project was instituted to promote the Central Highlands' economic growth and the development of the ethnic minority groups within the region. Since it was a national-level project, its decision-making process was typically top-down. In 1990, the Ministry of Energy prepared a technical analysis of the project and this was approved by the National

Box 9.1 Yali hydropower project – some basic data

Project cost: US$1 billion
Reservoir area: 6,450 ha
Reservoir capacity: $1,073 m^3$
Design capacity: 720 MW (3.6 billion kWh per year)
Number of displaced people: 6,782
Agricultural land flooded: 1,933 ha
Rice land flooded: 1,333 ha
Forest flooded: 3,492 ha

(NIAPP, 1993)

Appraisal Committee. On 24 September 1992, the Prime Minister formally approved the project and highlighted its purpose of promoting economic growth in the region and providing electricity for national development. The project's construction began on 4 November 1993. The Ministry of Energy, through its subsidiary Vietnam Electricity (EVN), played the key role in preparing as well as implementing this project. Theoretically, the peoples' committees at the provincial and district levels, as well as all of the line agencies and mass organizations such as the Youth Union and Fatherland Front, should also have been involved in this process. Since the project was in the national pipeline and had already been approved by the Prime Minister, however, the involvement of other agencies and interest groups was mostly only procedural. Some interviewees at the provincial level were not even aware of the project's EIA. Our interviews with the resettled communities revealed that people barely knew about the project until they had to be resettled. There were no environmental groups working on dam and river protection issues at that time, which also meant that civil society had no voice regarding this project's planning and construction.

In brief, for the Yali Falls Dam project, the hydropower narrative was unchanged from that of the 1970s and 1980s, and decisions involving the project were made in ways similar to those regarding other hydropower projects before that time, such as the Hoa Binh Dam. People were not consulted and civil society was not involved in the process. This was the period of an uncontested narrative on dams. Information was contained. During the 1990s, almost no one knew about the negative impacts of the dam, except for people in the affected area.

The Pleikrong Dam

Ten years after the Yali Falls Dam, the Pleikrong Dam was built. The dam was approved by the Prime Minister on 22 June 2001. Construction started in November 2003. Since the project was also in the national plan and implemented by EVN, the decision-making process of the project was similar

Box 9.2 Pleikrong hydropower project – some basic data

Project cost: US$204 million
Reservoir area: 53.28 km² (5238 ha)
Reservoir capacity: 1,048.7 million m³
Design capacity: 100 MW (478.5 GW per year)
Number of displaced people: 4,537
Agricultural land flooded (for other crops): 3,556 ha
Rice land flooded: 77 ha
Forest flooded: 226 ha

(Decision No676-QD-TTg)[9]

to that of the Yali Falls Dam, which means that everything was top-down. There was no opportunity to question the need for the project, as it was included in the energy development strategy for the period 2001–10. Hydropower was highly prioritized during that time, and NGO involvement in river protection was still limited. Our interviews show that, as with the Yali Falls Dam, government officials at the provincial level were not even aware of the project's EIA.

Benefits trickling down to the people at the dam site?

Hydropower in the Central Highlands contributes approximately 20 per cent of the national grid's electricity production annually, which is equivalent to a contribution of VND 1,920 billion (around US$96 million) to the national budget.[10] While dams may bring benefits to lowland and urban people, they often impoverish people who live in the dams' vicinity and they also damage the environment. Dam-affected people – mostly upland ethnic minorities – have suffered the most from the loss of land and assets due to inundation and resettlement. Overall, according to a report by the MOIT (2013), hydropower projects in the Central Highlands have affected 25,000 households. Thousands of hectares of land as well as houses, schools, clinics and offices are now under water due to dam flooding. At the national level, hydropower development inundated more than 65,000 hectares of land, of which 26,000 hectares were forest (MOIT, 2013). According to the Environmental Protection Law, projects that destroy forest must replace it, but currently only about 3 per cent of the forest lost has been replaced, with most of it being monoculture and subsequently having very low biodiversity (MOIT, 2013). In our interviews, the main reason given for this was that there is no land available for reforestation. Dams have also flooded large areas of agricultural land, resulting in persistent land shortages for the resettled communities.

For this research, we made several trips to Ho Moong and Sa Binh communes, Sa Thay district (Kon Tum province), where there are resettlement

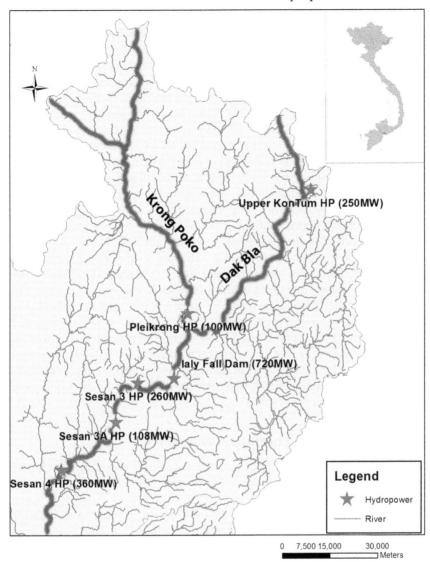

Figure 9.2 Sesan hydropower cascade

Source: Ialy Hydropower Company.

communities from both the Yali Falls Dam and Pleikrong projects (Figure 9.2). In 2006, more than 3,800 people in 748 households in the districts of Dak Ha, Dak To, Sa Thay and Kon Tum city had to move to resettlement sites in Ho Moong commune (Sa Thay district) in order to make way for the Pleikrong project. Our interviews show that the affected people were not consulted before the dams were built. They were just informed about the projects and asked to move to the resettlement sites. Seven years later, people are still struggling with their livelihoods due to the lack of farmland. Before the dams were built, each household owned on average 6–7 hectares. After the dams were built, each household of five people received one hectare of farming land; households with more than five people received 1.2 hectares. Since the planners did not include contingent land in preparing the resettlement site, there was no extra land for population growth. Seven years after the resettlement, the families have expanded and the number of families has increased but there is no additional land for them. As a consequence, in Ho Moong commune, 66 per cent of the residents are ranked as 'poor' (Ho Moong People's Committee, 2012), while in Sa Binh commune this figure is 43 per cent (Sa Binh People's Committee, 2012). These levels of poverty can be directly attributed to resettlement as many of them do not have enough land for farming, and as a consequence they face food insecurity and are becoming impoverished.

There are many dam-associated issues that have not been easy to deal with, in particular the livelihoods of local communities. At the 2012 year-end meeting of the Central Highlands' Steering Committee, the issue of hydropower development was once again intensively discussed. Besides the issues of deforestation and resettlement, there were also concerns about the destruction of cultural landscapes, the ecological environment and degradation of the water resources in the area, such as the problems at the Gia Long and Dray Nur waterfalls in Buon Ma Thuot, which dried up due to upstream hydropower water use. So far there have been no policies to effectively address these problems. There continue to be social, environmental and economic risks associated with dam building. This has led to changes in the way people view hydropower in Vietnam.

The impact of shifting narratives and policy

The emergence of critical hydropower narratives spearheaded by NGOs and academics has changed the landscape of hydropower development in Vietnam and its related policies.

For the Yali Falls Dam, EVN was the investor. It was responsible both for constructing the project and implementing the resettlement plan. The policies that governed the resettlement and compensation for land loss were nascent. Most of these policies were issued straight after the 1993 Land Law, and Yali Falls was the first project to apply these resettlement policies. The level and scope of the compensation were limited compared to later policies. The conditions for compensation were quite strict. For example, Decree 90 CP, dated 17 August 1994, emphasized that only people who had legal documents to their land were allowed to receive compensation for land loss when the state appropriated land for the purpose of national security, national interest or public interest, as defined in Article 27 of the 1993 Land Law. People who did not have these legal documents received little or nothing from the project. The same was true of ethnic minorities who did not 'own' land per se, but practised swidden agriculture in the Central Highlands forests.

By the time the Pleikrong project started, resettlement policies had improved somewhat. Decree No. 197(2004)/ND-CP on guidance for compensation, assistance and resettlement when appropriating the land of affected households, for example, contained more detailed instructions for resettlement implementation. The new decree separated out the cost of compensation from the cost of construction at resettlement site, and also loosened the conditions for compensation. As long as the affected people lived on the land, their ownership was not disputed and they received compensation. The decree also gave more detailed instructions on supporting displaced people in redirecting their livelihood activities. In contrast to the Yali Falls Dam, implementation of the resettlement programme was no longer the responsibility of the investor, but was transferred to the local peoples' committees, who were responsible for resettlement including land recovery, compensation and livelihood rehabilitation.

Despite these improvements, the policy changes have also led to increased problems with licensing new projects, implementation and impact mitigation. When the government issued new policies to promote faster hydropower development, a large number of hydropower projects were licensed within a short time, 'before the local authorities and people realized the dangerous devastation of the plants on the land and local people's lives' (Vietnamnet, 2013a). For example, Decision No. 797/CP-CN, issued in 2003,[11] gave more power to investors, aimed at speeding up construction and quickly getting more hydropower projects implemented. Pleikrong was the first project to apply a special mechanism that allowed the investors (EVN) to select contractors without an open bidding process. This new policy, according to some Central Highlands governmental officials, has created favourable conditions for cost distortion, potential favouritism and political influence.[12] The policy has also resulted in lower project cost-effectiveness, dam safety and more negative impacts on local communities and the environment (Vietnamnet, 2013b). If dam builders have more power, problems of corruption (especially for state-owned projects) and construction quality may arise more frequently. As mentioned above, evidence of these issues already exists – the Pleikrong Dam was reported to be overspent and the Dak Mek 3 and La Krel 2 dams broke in 2012 and 2013.

These new policies, which established the concessional conditions for hydropower development, have resulted in a shift in how power is exercised in development. For example, investors have more power in decision-making processes concerning project size, design and construction. For the Pleikrong project, since EVN was allowed to control the construction and decision-making process, it changed the project design without consulting any other agencies. Specifically, in a document approved by the Prime Minister,[13] the installed capacity of the project was 110 MW, while in all of the other technical documents later approved by the Ministry of Industry the installed capacity was listed as 100 MW. This change was not formally approved by the Prime Minister (Baomoi, 2009). There were also problems with resettlement due to the project's design. For example, since communities in Ho Moong and Sa Binh communes had to resettle twice, money was wasted on building new roads, houses and infrastructure which were later submerged under the water. Deforestation in the area became more serious as people did not have enough land; they had no choice but to clear the forest for their livelihoods. On a positive note, these negative impacts attracted media coverage and have led to more research on dams and their impacts.

A number of other issues related to the Pleikrong Dam have emerged in the media, especially issues related to its approval processes. First, there was a delay in budget approval. Although the project started on November 2003, its total estimated budget was not approved until 2006, three years later. This budget approval delay contravened the policy which requires investors to complete budget approval no later than six months after approval of the technical designs (Decision No. 797/CP-CN, 2003). Furthermore, construction unit prices were approved in 2007, more than a year after the total budget was approved. This meant that EVN violated the law by approving the total budget before approving construction unit prices. EVN even approved the construction norms and unit prices before getting feedback from the Ministry of Construction. Although EVN wielded power during the project design and construction, some of the project expenditure did not follow the rules. For example, Song Da joint venture consulting company in the Central Region claimed 8 per cent of the consulting work for preparing a detailed cost estimation, although in fact they did not actually perform this work. EVN approved an amount of VND 488,167,000 (US$31,500 in 2006) for this 'unreal' cost (Baomoi, 2009).

The changing flow in hydropower narratives has occurred not only within civil society, but also among government officials at various levels, from national down to commune level. Even central government has begun to signal its growing realization that it may be necessary to reduce its reliance on hydropower. Since 2007, an increasing number of the National Assembly (NA) members have been concerned about the rapid development of hydropower in the country. They have raised many questions about dam impacts in NA meetings (Dao, 2012). As a result, in late 2009, the Prime Minister required the MOIT to prepare an overall assessment of the country's hydropower situation. Released in 2010, the MOIT's assessment suggested cancelling 38 small and medium hydropower projects due to their potential negative impacts, and it recommended reconsideration of 35 others before commencing construction (MOIT, 2010). It is worth noting that all of these projects are small and their investors are mostly from the private sector. As mentioned earlier, large projects are included in the national power planning and the majority of them had already been built or were under construction by state-owned companies. Therefore, this outcome, even though positive for river protection, is in fact not free from criticism. Some argue that it is not truly a result of narrative or political change, but is simply a move in an economic game in order to protect big projects and appease public opinion.[14] Whatever the real reason behind it and even though this number was insignificant compared to the number of dams already in the planning stages, it shows how concern about hydropower overdevelopment and its associated problems has emerged within the state.

In 2012, the NA continued to require that the MOIT, together with other related ministries and provincial people's committees (PPCs), review hydropower projects, both ongoing and under study (Resolution No. 40(2012)/QH13). Dr Pham Hong Giang, chairman of Vietnam's Large Dam Association, and former Vice Minister of Agriculture and Rural Development, was quoted in *TienPhong* newspaper on 10 April 2013, as saying that:

> Our recent hydropower development has shown that we have not thoroughly studied these issues. First, hydropower planning was done in a fragmented way without taking into account the issue of integrated water management in order to achieve the highest socio-economic effectiveness in a basin; second, we neglected and even ignored the impacts of changes in hydrological flow on downstream environments and communities; third, our cadres have been incompetent and irresponsible in planning and developing hydropower.
>
> (VNCOLD, 2013)

Even the *Nhandan* (*People*) newspaper, the formal voice of the Communist Party, recently expressed alarm over the uncontrolled hydropower development and its safety in Vietnam, in particular, in the Central Highlands (*Nhandan*, 2013). According to the Central Highlands' Steering Committee, hydropower development in the Central Highlands over the last decade had led to many negative consequences, especially in regards to insufficient compensation for the affected people and environmental destruction. The Central Highlands' Steering Committee was aware of this problem and had been working with the MOIT to review hydropower development planning. It has suggested that 'inefficient' projects should be cancelled.

For example, Tran Viet Hung, a vice chairman of the Central Highlands' Steering Committee, was quoted in *Nhandan* as stating:

> The MOIT has completed the fieldwork for reviewing some hydropower projects in the region. The opinion of the Committee in this matter is that we have agreed to seriously review some hydropower projects in order to produce a comprehensive overall assessment of the problems. Those projects that might cause significant impacts on the environment and local people's livelihoods and do not have effective mitigation measures, we will definitely cancel. From now on, when we have not had a formal response regarding hydropower [from the NA], we should not start any new projects.
>
> (*Nhandan*, 2013)

As a result of this change in the hydropower perspective at a higher level, the MOIT has agreed with the PPCs to review all remaining projects in the pipeline (with a focus on small projects). In January 2013, after a careful review of projects' cost-efficiency and socio-environmental impact assessments, it was proposed that 67 planned projects and three potential projects be cancelled (total of 157.9 MW). Another 117 projects (total of 769.9 MW) are under strict review and will only be allowed to commence after 2015 if the study shows that the socio-environmental negative impacts of these projects are not significant to the locality (MOIT, 2013). By the end of 2013, another breakthrough occurred for social and environmental advocacy in Vietnam. The government announced its decision to cancel two projects, Dong Nai 6 (135 MW) and Dong Nai 6A (106 MW) on the Dong Nai River, which, if built, would seriously impact the Cat Tien National Park. These two projects were removed from the National Hydropower Project Development Plan (Document 1858/VPCP-KTN, dated 23 September 2013). The cancellation of these projects was the outcome of an extensive three-year campaign, between 2011 and 2013, by the VRN and many other stakeholders throughout the country. These projects are the first large hydropower projects to be cancelled due to the pressure of public opinion within the country. This may signify a significant shift away from destructive hydropower development.

Conclusion

The analysis above demonstrates that, even though decision-making in dam development in Vietnam is a top-down process, there has been a clear shift in hydropower narratives over time. The dominant hydropower narrative in Vietnam has been constructed by the state and its elite. For a long time, the narrative seemed to be uncontested. Information about dam-associated problems was contained and only known about in the affected areas. The affected people assumed that their sacrifice was for the 'national interest'. But the growing attention from NGOs, researchers and the media has revealed these problems to the wider public, which has led to a counter-narrative on hydropower that in turn has resulted in policy change. Not all of these policy changes have had positive impacts. A policy that favours investors can attract more problems, causing additional research and media attention on the topic, and putting additional pressure on the existing narrative.

The dominant perspectives on hydropower remain largely unchanged, despite the critical view of hydropower of some governmental officials. The recent report by the NA's Committee on Science, Technology and the Environment, while touching on some of the problems caused by hydropower, still emphasizes that hydropower is a clean and renewable energy source which helps ensure national energy security. More importantly, it highlights that hydropower has significantly contributed to promoting the socio-economic development of the country in general and of the upland areas in particular, where these projects are located (MOIT, 2013). The Revised Water Law 2012 still highlights the importance of hydropower and encourages hydropower development where possible on a nationwide scale.

Criticisms of hydropower development by NGOs, CSOs, the media and the public over the last decade counter the government's dominant narrative, but have not yet affected the way in which hydropower decisions are made. Even though land and resources policies have improved, decision-making on dam building still takes inadequate consideration of costs into account. In addition, since both the NGOs and the media are still strictly controlled by the government, there are certain issues which are not thoroughly discussed. The influence of NGOs and the media, even though they have increased over the last few years, is still limited.

There has been no change in how large hydropower projects are evaluated once they are part of the national planning process. In the cases of the Yali Falls and Pleikrong Dams, as elaborated above, the decision-making process was entirely top-down, even though the social contexts were different. By the early 2000s, there was greater awareness of the impacts caused by hydropower projects compared to the early 1990s. For the Pleikrong project, the situation was even worse because the investor was allowed to have more power. Despite this, there were still policy adjustments aimed at promoting hydropower development at maximum speed. Specifically, the MOIT gave permission to provincial governments to adjust hydropower project designs and implementation, and to provide supplementary planning. This policy enabled hydropower plans to be amended easily, mostly in favour of investors.[15] Even though government policy requires discussion among the relevant authorities, departments and agencies at all levels during the hydropower planning process, this has only been a superficial process. In fact, the current regulations for decision-making have greatly empowered investors after they have received investment approval. Investors have full power to decide on the technical design and to monitor the quality of construction and the project's execution. The regulations allowing adjustments to planning have facilitated improper planning implementation and have created gaps enabling investors to bend the rules, as in the case of the Pleikrong Dam.

Although it seems that the changing narrative has not been helpful to the way in which decisions are made, the recent governmental review of hydropower development is considered by the public to be a positive move in addressing the public outcry over the negative hydropower impacts. Both central and local government have attempted to address these problems, and the emerging critical counter-narratives from environmental groups and the public have shed new light on hydropower politics and related decision-making. Indeed, research and publications by non-governmental bodies have helped the public to gain a more comprehensive understanding of hydropower issues and have led to a push for changes to hydropower policies and their implementation. Above all, the last few years of campaigning by NGOs has changed public perspectives on hydropower and awakened the population to its impacts on the uplands. It is now more widely recognized that this development can no longer be seen as a central way of modernizing highland populations.

Notes

1 Before 1975, the country was divided into North and South Vietnam. Thus, the Vietnamese state here means the northern government.
2 Interview with official from the Company of Power Investigation and Design No. 1 in Hanoi on 10 May 2013.
3 Outsiders may see civil society in Vietnam as a strange situation, in which VUSTA, on the one hand, is formally a state organization and, on the other hand, is an umbrella organization, which controls and regulates hundreds of non-government organizations in Vietnam. Scientists who work under VUSTA, to some extent, can be seen as part of civil society. In the case of the Son La Dam, opinions from scientists under VUSTA were listened to by the government.
4 In Vietnam, by law, networks are not allowed to stand by themselves. A network must be hosted by a registered organization and follow the organization's operating regulations. That has caused difficulties for the network to operate.
5 VTV2 (a national TV channel) often covers news about the environment that includes water and food security, river issues, hydropower and resettlement, etc. *Tuoitrenews, Thanhnien, Tienphong* are several of the newspapers which cover many stories related to hydropower and its associated problems including resettlement and deforestation.
6 From 1990 to 2000, 160,000 households with about 810,000 people spontaneously migrated from the lowlands to the Central Highlands. Most of the migrants were of the Kinh ethnic group; they accounted for 64 per cent of the total (Vietrade, 2011).
7 In some Vietnamese documents, the Yali Falls Dam is spelled as Ially or Yaly.
8 Meeting with Sa Thay district authorities on 16 May 2013.
9 The data come from the formal website of the Legal Library, http://thuvienphapluat. vn/archive/Quyet-dinh-676-QD-TTg-dau-tu-du-an-thuy-dien-Pleikrong-vb177139.aspx.
10 Data added up from individual project reports.
11 Decision No. 797/CP-CN issued in 2003 on hydropower investment policy.
12 Interviews with local officials in Kontum on 27 May 2013.
13 Decision No. 676/QĐ-TTg.
14 Interviews with officials from Ministry of Agriculture and Rural Development (MARD) on 10 May 2013.
15 Article 14 of Decision No. 55(2008)/QD-BCT dated 30 Dec. 2008.

References

Baomoi (2009) 'Pleikrong Dam – mistakes everywhere' www.baomoi.com/Du-an-Thuy-dien-Pleikrong-tinh-Kon-Tum-Dung-den-dau-sai-den-do-Ky-1-Thi-cong-rua-bo/148(3173671).epi, 6 Sept., accessed April 2013.

CAND Portal (2012) 'Hundreds of families got into difficult situations due to hydropower' www.baomoi.com/Source/CAND-Portal/, accessed 30 July 2014

CEMA (Committee on Ethnic Minority Affairs) (2011) *Population Statistics by Ethnicity*, Hanoi: CEMA.

Chu, T., Nguyen, T. H., and Truong, X. P. (2007) *Social Impacts of the Tuyên Quang Dam on Local People*, Hanoi: Vietnam Union of Science and Technology Associations.

CRES (Centre for Natural Resources and Environmental Studies) (2001) *Study into the Impact of the Yali Falls Dam on Resettled and Downstream Communities*, Hanoi: CRES and Vietnam National University.

Dao, N. (2010) 'Dam development in Vietnam: The evolution of dam-induced resettlement policy', *Water Alternatives,* 3(2): 324–40.

Dao, N. (2011) 'Damming rivers in Vietnam: A lesson learned in the Tây Bắc (Northwest) Region', *Journal for Vietnamese Studies*, 6(2): 106–40.

Dao, N. (2012) 'Displacement, resettlement and agrarian change in the Northern Uplands of Vietnam', PhD thesis, York University, Toronto.

Dao, X. H. (2009) 'Irrigation in Vietnam: Achievements and challenges in development', opening speech at 63rd Ceremony of Traditional Water Conservation Day at the Hanoi Water Resources University, 28 Aug.

Decision 676-QD-TTg on Pleikrong Dam (2002) http://thuvienphapluat.vn/archive/Quyet-dinh-676-QD-TTg-dau-tu-du-an-thuy-dien-Pleikrong-vb177139.aspx, 15 Aug., accessed June 2013.

Decision 797/CP-CN/2003 (2003) http://thuvienphapluat.vn/archive/Cong-van/Cong-van-797-CP-CN-cac-du-an-dien-khoi-cong-nam-2003-2004-vb28012t3.aspx, 17 June, accessed June 2013.

Diep, D. H. (1992) 'The characteristics of ethnic communities: Social development of the northwest area of Vietnam under the impact of dam projects', Working Paper, Hanoi: Institute of Ethnology (Original: Diệp Đình Hoa và tập thể tác giả, 1992, Những đặc điểm về cộng đồng dân tộc – phát triển xã hội Tây bắc Việt Nam dưới sự tác động của các công trình thuy điện, HàNội).

Diep, D. H. (1997) *The Transformation of Ethnic Communities: The Impact of HoaBinh Reservoir*, Hanoi: Social Sciences Publishing House (Original: Diệp Đình Hoa, 1997, Sự biến đổi của cộng đồng dân tộc. Tác động của hồ Hoà Bình. Nxb. Khoa học xã hội, Hà Nội).

Doan, B., and Nguyen, D. A. (2006) 'Assessment on life quality of resettlers in Ta Trach hydropower project, ThuaThien Hue province', presentation to Vietnam River Network's annual meeting in Hanoi, 5 Dec. (Original: Đánh giá chất lượng cuộc sống của người dân vùng tái định cư dự án Hồ Tả Trạch – tỉnh Thừa Thiên – Huế. Báo cáo khoa học, Hội thảo Vietnam River Network, Hà Nội, ngày 5(12)/2006).

Duong, A. T. (2009) 'Assessment of the consultation process with affected people in Bắc Hà hydropower project, Lào Cai Province', presentation at the VRN annual meeting, 27–9 Nov., Hải Phong.

Fairhead, J., and Leach, M. (1995) 'False forest history, complicit social analysis: Rethinking some West African environmental narratives', *World Development*, 23(6): 1023–35.

Government Portal (2013) 'Vietnam's socio-economic development strategies up to 2020', http://chinhphu.vn/portal/page/portal/chinhphu/noidungchienluocpha ttrienkinhtexahoi?_piref135_16002_135_15999_15999.strutsAction = ViewDetailAction.do&_piref135_16002_135_15999_15999.docid=654&_ piref135_16002_135_15999_15999.substract, accessed June 2013.

Hardy, A. (2002) *Red Hills: Migrants and the State in the Highlands of Vietnam*, Honolulu: University of Hawai'i Press.

Hickey, G. C. (1982) *Sons of the Mountains: Ethnohistory of the Vietnamese Central Highlands to 1954*, New Haven, CT: Yale University Press.

Hirsch, P., Bach, T. S., Nguyen, T. H. V., Do, T. H., Nguyen, Q. H., Tran, N. N., Nguyen, V. T., and Vu, Q. T. (1992) *Social and Environmental Implications of Resource Development in Vietnam: The Case of Hoa Binh Reservoir*, RIAP Occasional Paper 17, Sydney: Research Institute for Asia and the Pacific, University of Sydney.

Ho Moong People's Committee (2012) *Annual Report 2012.*

Hoang, L. A. (2006) 'Assessing life quality and potential for rehabilitation of the resettlers in the A Vuong hydropower project', presentation to the Vietnam River Network's annual meeting, 5 Dec., Hanoi.

Hoang, N. G., and Vo, H. C. (2006) 'Study on socioeconomic environment of resettlement area of Thác Bà hydropower project, Yên Bái, 32 years after the dam's construction', presentation to the Vietnam River Network's annual meeting, 5 Dec., Hanoi.

Kerkvliet, B. J. T. (2009) 'Everyday politics in peasant societies (and ours)', *Journal of Peasant Studies*, 36(1): 227–43.

MOIT (Ministry of Industry and Trade) (2010) 28/BC-BCT: Report to the Prime Minister Regarding the Hydropower Situation, internal document, Hanoi: MOIT.

MOIT (Ministry of Industry and Trade) (2013) National Assembly's Committee on Science, Technology and Environment's review on planning, investment, construction and operation of hydropower at nationwide level, 24 April, internal document, Hanoi: MOIT.

National Hydropower Plan Study (1998) *Final Draft Report, Volume VII: Sesan River Basin*, Hanoi: SWECO International, Statkraft Engineering Norplan.

Nhandan (2013) 'Alarming about safety of hydropower projects' (Báo động về an toàn các công trình thủy điện), www.nhandan.com.vn/mobile/_mobile_xahoi/_mobile_ tintucxh/item/20559102.html, 15 June, accessed July 2013.

NIAPP (National Institute for Agricultural Planning and Projection) (1993) *Resettlement and Rehabilitation Programme for the Yali Falls Dam Project*, Hanoi: NIAPP.

Nguyen, D. O. (2009) 'Hydropower development in Vietnam: Potential, existing conditions, and development plan', paper presented at a training workshop on World Commission on Dams' Strategies, held by the Vietnam Rivers Network on 7–8 Nov., Hoa Binh Province, Vietnam.

Nguyen Tan Dung (2012) http://baodientu.chinhphu.vn/Home/Khanh-thanh-cong-trinh-Thuy-dien-Son-La/201212(157501).vgp, 23 Dec., accessed June 2013.

PhapLuatTPHCM (2011a) 'Hydropower in the Central Highlands and its associated problems – deforestation for hydropower'www.baomoi.com/Thuy-dien-Tay-Nguyen-va-he-luy-Bai-1-Pha-rung-lam-thuy-dien/148(7435635).epi, 28 Nov., accessed 25 April 2013.

PhapLuatTPHCM (2011b) 'Dry river – thirsty people', http://phapluattp.vn/ 20111128111347387p0c1085/thuy-dien-tay-nguyen-va-he-luy-bai-2-song-kho-dan-khat.htm, 29 Nov., accessed April 2013.

PhapLuatTPHCM (2011c) 'Gave land to hydropower and went hungry' http:// phapluattp.vn/2011112910101549p0c1085/thuy-dien-tay-nguyen-va-he-luy-bai-3-nhuong-dat-cho-thuy-dien-roi-doi.htm, 30 Nov., accessed April 2013.

Power Development Plan No. 7 (2011) 'Decision No. 1208/QD-TTg of July 21, 2011, approving the national master plan for power development in the 2011–2020 period, with considerations to 2030' www.nti.org/media/pdfs/VietnamPowerDevelopmentPlan2030. pdf?_=1333146022, accessed April 2013.

Reisner, M. and Bates, S. (1990) *Overtapped Oasis: Reform or Revolution for Western Water*, Washington, DC: Island Press.

Sa Binh People's Committee (2012) *Annual Report, 2012*.

Salemink, O. (2003) *The Ethnography of Vietnam's Central Highlanders: A Historical Contextualization, 1850–1990*, Honolulu: University of Hawai'i Press.

Socialist Republic of Vietnam (1992) Constitution of the Socialist Republic of Vietnam, Hanoi: Socialist Republic of Vietnam.

Son La People's Committee (2006) *Compiling of Policies on Resettlement for the Son La Hydropower Project*, Son La: Son La People's Committee.

Thanhnienonline (2013) 'Hydropower kills rivers', www.thanhnien.com.vn/pages/20131118/thuy-dien-giet-song.aspx, 18 Nov., accessed January 2014.

Trang, H. D. (1995) *Scientific Basis for Stabilizing and Rehabilitating Resettlers*, Hanoi: Ministry of Agriculture and Rural Development.

Vietnamnet (2013a) 'Vietnam repeats "hydropower mistake" with oil refineries?', http://english.vietnamnet.vn/fms/environment/87352/vietnam-repeats-hydropower-mistake-with-oil-refineries.html, 23 Oct., accessed October 2013.

Vietnamnet (2013b) 'How many hydropower projects are under careless construction?' (Còn bao nhiêu thủy điện thi công ẩu?), http://m.vietnamnet.vn/vn/xa-hoi/128036/con-bao-nhieu-thuy-dien-thi-cong-au.html, 21 June, accessed June 2013.

Vietrade (2011) www.vietrade.gov.vn/vung-kinh-te-tay-nguyen/2378-vi-tri-dia-ly-va-dieu-kien-tu-nhien-vung-kinh-te-tay-nguyen-phan-1.html, 3 Nov., accessed June 2013.

Vietrade (2012) www.vietrade.gov.vn/vung-kinh-te-tay-nguyen/3297-thy-in-trng-ca-mi-tren-tay-nguyen.html, 5 Dec., accessed June 2013.

VNCOLD (Vietnam Commission on Large Dams) (2013) www.vncold.vn/Web/Content.aspx?distid=3266, 10 April, accessed June 2013.

Vnexpress (2013) http://vnexpress.net/tin-tuc/xa-hoi/de-nghi-loai-bo-nhieu-du-an-thuy-dien-o-tay-nguyen-2853591.html, 22 July, accessed July 2013.

Vo, V. H. (2006) 'Study on the environmental and social situation at resettled areas of Bản Vẽ hydropower project, Nghệ An', presentation to the Vietnam Rivers Network annual meeting, 5 Dec., Hanoi.

VoV (2013) 'Medium and small hydropower in the Central Highlands – Stop before it will be too late' http://vov.vn/Xa-hoi/Thuy-dien-vua-va-nho-o-Tay-Nguyen-Dung-truoc-khi-qua-muon/250109.vov, 28 Feb., accessed April 2013.

VRN (Vietnam Rivers Network) (2011) *Studies on Hydropower Development Induced Resettlement in Vietnam during 'Doi Moi'* (Nghiên cứu tái định cư thuy điện ở Việt Nam thời kỳ đổi mới], Nhà xuất bản Từ điển Bách khoa), Hanoi: VRN.

VUSTA (Vietnam Union of Science and Technology Associations) (2006) *A Work in Progress: Study on the Impacts of Vietnam's Son La Hydropower Project,* Hanoi: Vietnam Union of Science and Technology Associations.

Index

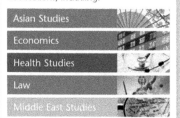
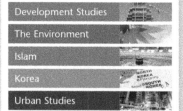

Milton Keynes UK
Ingram Content Group UK Ltd.
UKHW031149141024
449569UK00024B/949

9 781138 377509